THE EXPERT WITNESS

A MANUAL FOR EXPERTS

THE EXPERT WITNESS

A MANUAL FOR EXPERTS

BY

ROBERT J. CRAWFORD, P.E.

CONSULTING CIVIL & STRUCTURAL ENGINEER

1stBooks – rev. 1/25/01

TABLE OF CONTENTS

TITLE   **PAGE**

ACKNOWLEDGEMENTS

My sincere appreciation is given to my good friend Judge John M. Bodley for his assistance in reviewing both the legal and grammatical presentations in the text of this book. Similarly, appreciation is given to Rae Cochrane for her patient editing assistance and encouragement in ensuring my words herein will be understood by those who read them. And to my wife, Maureen, my heartfelt appreciation for her patience and understanding of the many hours it has taken to present my thoughts in the pages of this book. Cover design by Raschelle Herbolsheimer.

PREFACE

This book is a practical guide to provide education and direction to those who come into contact with the litigation area of the law, either as a result of one's personal involvement or as requested by others in need of expert assistance. Its mission, with respect to the expert witness, is to explain how the legal system operates, the functions of those who practice therein, and to teach the methods found to be essential for successfully engaging in this line of work. While the field of the author is construction, the approach to problems and techniques applies equally well to all walks of life be they in medicine, accounting, business, insurance, or any other field of endeavor. The procedures of investigation, information collecting, report writing, and testimonial participation in the legal arenas of mediation, arbitration and courts of law are essentially the same. Working with people, especially the lawyers, is both interesting and challenging, and the skills required are independent of one's field of expertise; human beings follow similar patterns of behavior regardless of their personalities or fields of endeavor.

The information presented here is the culmination of more than thirty-five years of the author's personal experience in observing, developing and implementing the many techniques and procedures contained in the following chapters. To enter the legal arena requires the participant to be prepared; the best way to do this is to learn how it works, how to best utilize one's talents, and to be thoroughly prepared in his field of expertise for the undertaking.

INTRODUCTION

America's society has become one of a litigious attitude. The overabundance of practicing lawyers in the marketplace has given easy access to the courts for claims that formerly would have been literally laughed about – and that few responsible attorneys would even consider. What would have been generally settled with a few words and a handshake just a few years ago now finds one of the participants in a dispute running to an attorney to try to capitalize on the situation – and, there is always some enterprising attorney eager to join the fray. Recent statistics rumor there is an attorney for every eight people in California, and they are diligently working on the remainder of the United States. Small wonder the increase in lawsuits – especially the frivolous variety.

Today, people in all walks of life, especially those in business, have become unsuspecting targets for those individuals who seemingly have discovered an easy method for generating income. Indeed, there are some who find a livelihood can be had simply by suing others about the most trivial of causes. When an attack is made on an unsuspecting person – or business - the recipient finds himself in the position of having to defend against it regardless of the claim. The only prudent action to be taken is to hire a lawyer for his defense. Lawyers are expensive, and defense against so-called "frivolous" lawsuits often ends up in settlement conferences because of the high cost of litigation and court delays.

If the recipient has insurance, he will normally file a claim with his carrier. The agent representing the carrier checks to see if the client's policy covers the claim. If it does, the agent's task is to minimize costs to the company. If it is not covered, he so informs the policy owner. When an insurance company does become involved, their concern in the problem centers on the probable cost rather than the issue of who may be right or wrong. While their direction of effort ultimately points toward litigation, that avenue is used only if they cannot obtain agreement to a lesser cost through settlement. Plaintiffs in such cases normally have no out-of-pocket expenses as their attorneys usually work on a percentage of the proceeds, if any. If the plaintiff is successful, he and his attorney walk away wealthier than when they started; if he loses, he is out nothing and the attorney is out only the time he spent on the case. These so-called professional suit-chasing individuals don't win very often, but they are always on the lookout for an easy target.

Obviously, not all claims are frivolous. There are those that indeed have justification coming from both sides of an issue. Such claims contain a true difference of opinion that obviously requires a judicial decision. It is the task of the attorneys representing each side to prosecute or defend in hopes of convincing the court of justification for their side of the issue. Attorneys know the law. They know the procedures and courtroom tactics to be used in presenting their cases to the court and jury. What they don't know nor fully understand – and often need assistance with – are the essential technical facts of their cases. For this, they turn to others. This is the realm of the expert witness. Experts are used to assist attorneys in all fields of business and human relationships. Wherever there is a court trial, there are usually experts giving testimony on the issues under consideration. There are experts in every field of endeavor; they come in all sizes, shapes, personalities, and degrees of knowledge and experience. Some are very good, some not so good; it is the attorney's task to try to select experts who are knowledgeable and experienced both in their field and legal procedures – not an easy job.

In my years of professional practice, I have served many times as an expert witness. Throughout this period, I have observed the actions of numerous other experts – some on my side as well as those in the opposing camp. One thing that has dramatically stood out is how many of them have little or no idea of how to approach their tasks. I have seen qualified people make damaging or fatal mistakes simply because they did not know how to organize their information to present it properly to their clients, or in courtroom testimony. A good deal of the problem is just unfamiliarity with the legal process, with how to present themselves properly, and with the ability to present their knowledge and experience in focusing on an issue. The expert should be at ease with the legal process and knowledgeable of the attorney's tasks. He should understand what the attorney is trying to accomplish as well and the direction the attorney is taking – so he may be of real assistance, not just a dictionary reciting definitions. He should be able to coach the attorney so that they may work together as a team for presentation of testimony. He should also be able to think clearly!

This book was written to present guidelines for the person who has been voluntarily – or involuntarily – called upon to serve as an expert witness whether for himself or others. It is also intended to educate attorneys on how to select and get the most from their experts in this manner.

During my years of observation and participation, the legal arena has appeared essentially as a game. It is a deadly serious game generally consisting of high stakes where attorneys apply a highly organized and determined effort in preparation and presentation of their cases. During court proceedings, personal posturing is often used and is part of the plan in presenting information to influence a jury. Attorneys are good actors and most play their roles well. It is an interesting experience to work with a capable attorney during the testimony phase of a trial.

The information contained in the following chapters was gleaned from personal observation and experience during service as an expert witness. Each chapter is presented progressively in the process of practicing this interesting and challenging art/science form. Each chapter covers a specific portion of the whole field but is not dependent upon the other chapters for understanding. Some information, however, is closely related and is briefly touched upon to ensure continuity between various chapters.

The path followed in using one's knowledge and experience while serving a client in the legal arena contains many rewards in both personal satisfaction and remuneration. But it can also be fraught with pitfalls to be avoided. Having a map to follow when the territory becomes unfamiliar is both necessary and reassuring. Serving as an expert witness is a combination of intelligent application of knowledge, a good memory, and an awareness of emotion while working with others.

Chapter 1 explores the areas of our society in need of technical assistance. Who are the people that require the services of experts? Its purpose is to acquaint the reader with those who use experts both occasionally and on a regular basis in their work. It elaborates on their businesses, personalities, and what to expect when working with them. It also provides advice when called for assistance as to what inquiries should be made about their businesses, reference checking, and the essential elements of the Agreement for the work. Because these "clients" are the source of business for experts, it pays to know when to accept an assignment - and when to refuse.

Chapter 2 describes requirements the practicing expert should possess. It delves into education, experience, and related capabilities in his field of practice – whether he is a professional person or skilled craftsman. It also is essential that the expert be capable of oral and written presentation of his thoughts. Because the expert is a professional person in his field, acceptable personal characteristics in speech, dress, vocabulary, personality and personal conduct are a must for credible posturing while testifying in deposition and court proceedings. An expert should look and act as an expert. Additionally, he should be capable of analyzing opposing experts for strengths and weaknesses that may aid his clients in the legal contests.

Chapter 3 is where the expert goes to work. This outlines the investigative efforts he makes on behalf of his client. It begins with the client interview where the essence of the problem is discussed, then moves on into document gathering for information and study purposes, visits to the location of the problem, and a multitude of other fact gathering chores. When all available information has been collected, the expert begins his study. This chapter guides the expert through the maze of fact finding duties pointing out the type of information that is needed for his analysis and the various barriers to be overcome during the collection process.

Chapter 4 addresses the task of placing relevant information gathered during the investigation in report form. It weaves the pattern of chronological events that led the expert through his study concluding with his findings and professional opinions in the case. It becomes his professional opinion as an expert witness and will stand in his place if he is unavailable for personal appearance. The chapter includes information on the construction of the report for easy understanding, including the use of acceptable language, grammar, and form layout - as well as descriptions and photographic data to make the report both interesting and informative.

Chapter 5 introduces the expert to the officers of the court – the attorneys. It is an overview of these interesting practitioners of the law outlining their education, background, experience in and method of approach to legal issues, and especially to expert witnesses – both theirs and the opposition's. It describes the various roles they play in their practice and as employers of expert witnesses. They are usually bright, intellectually challenging people with a flair for dramatics. The chapter addresses examples of personalities the expert will meet in the general attorney population. It describes working with attorneys in the legal system and presents an outline of effective ways the expert may be of assistance. It also explores some of the many tactics opposing attorneys use to discredit or impeach expert witnesses and how to handle such attacks.

Chapter 6 puts the expert on the firing line during the deposition proceedings. It describes the process from subpoena through conclusion of the discovery session. Depositions are the major tools attorneys use to elicit information from opposing expert witnesses. The chapter fully describes the process of how to conduct oneself during questioning, and most importantly, how to answer questions without becoming confused or put off guard. It also stresses the importance of remembering all testimonial information given during deposition; if the expert changes his opinion or makes a mis-statement while on the witness stand in court – without first informing his attorney client, it will lead him to grief.

Chapter 7 orchestrates the information of the expert witness when he is preparing for court testimony. Because trials are usually convened some two years or so after the time of the problem, information becomes hazy and difficult to remember. This chapter describes how to categorize and store information so it can be recalled without difficulty. It also describes the procedure of the courtroom to prepare the expert for his entry and testimony. It further describes the important working cooperation with his attorney for testimony presentation and includes research to prepare for the personality and tactics employed by the opposing attorneys.

Chapter 8 describes the various forms of legal proceedings. It explains the structure and authority of Mediations, Arbitration, and the Courts. It acquaints the expert with what to expect and how to conduct himself in each of these arenas. It gives guidelines for the question and answer format while working with his attorney, and how to conduct himself during the cross-examination phase where the opposing attorney's task is to discredit his testimony. A section is devoted to the expert's personal demeanor in telling his side of the story to the jury; it is the jury who determines the outcome of a case and it is extremely important they develop a favorable attitude toward the expert.

Chapter 9 outlines the often omitted task of reviewing the trial, its outcome, and the role the expert played in the process. It is a learning experience to review the plusses and minuses of his performance, how he could have done better, and what to avoid in the future. Review is also made of the attorney's performance and how he followed the prepared script developed between himself and his expert. A study of the opposing attorney's presentation, and that of his experts, should also be made to gain additional information. The object is to learn as much as possible to hone skills and to avoid mistakes made in the past performance. Finally, some attention is directed to the jury's reaction to the players in the case, especially the expert witnesses.

Chapter 10 addresses compensation for services as an expert witness. The various ways of being paid for services are discussed along with the formation of the original Agreement with the client. Also presented are the various methods of payment, bookkeeping requirements and, most importantly, the collection of fees are discussed as well as tactics for dealing with slow or late payers, as well as non-payers.

A Glossary of legal terms is presented as an appendix to the book. These terms are the most commonly used and often are not readily understood by those not versed in the language of the law. They are presented as an aid for quick reference when such terms are encountered.

As a final note, the field of law and those who practice therein as attorneys, experts, judges, etc. are of both the male and female gender. The use of the male gender in the text of this book is intended to represent both sexes equally.

CHAPTER 1: CLIENT RELATIONSHIPS

This chapter will cover the essence of the investigation business - the clients we serve. Because they are so important to the work we perform, we should know who they are and understand the manner in which their industry works.

Clients come in all sizes, shapes and attitudes and have one thing in common: problems. It is because of their needs for assistance that they turn to an "expert", someone who is supposed to have the answers to their questions or at least know how to get them. It is our job, as practicing professionals and skilled craftsmen, to offer such assistance as is within our knowledge and abilities. So the purpose of this chapter is to identify those who, from time to time, have need for our services; from this we will gain some insight into their personal characters and respective business operations. The reason for this exploration is to attain a knowledgeable appreciation of their approach to business and method of operations so we may be more aware of how to service their needs.

Our society has become so complex with its myriad of laws, regulations and rules that a real need has developed for people with specialized knowledge. With the vast increase in the number of attorneys, legal action in one form or another has increased dramatically. Lawsuits appear to develop readily where previously a problem would have been settled in a more mutually beneficial or agreeable manner. The rule for today seems to have become: life, liberty, and the pursuit of litigation. In all fairness, however, the attorneys wouldn't be there if the need didn't exist.

In those situations where experts are required, the view point of the client usually comes from one of two directions. He has either been wronged by some other party and desires to take action against that party or someone has taken the offensive against him and finds he must defend himself.

In the former case, the client would be considered the plaintiff; the latter, the defendant. There has been some discussion in the industry regarding the position the expert takes when he assists a plaintiff. Some attorneys try to create a disparaging attitude toward an expert when he represents the plaintiff - but in reality, it is only a ploy to psychologically discourage the expert from doing his best, painting him as assisting the "bad guy." In truth, whichever side the expert finds himself serving, there is a justifiable reason for the action and he should always remain unruffled, unbiased, and put forth his best efforts consistent with abilities.

This leads to a consideration of how the expert should be selecting his clients. All but a fortunate few actively need and seek new clients because they are the essence of our livelihood. But no matter how much we would like to accept all potential clients who approach us, some general guidelines are necessary for acceptance or rejection. In my practice, I've found the following to be a good list to refer to:

1. Try to get a good reading on the character and background of the client. Does he appear to be an honest and trustworthy person? If it's a sizable problem involving a considerable amount of money, check him out by asking for references; check his bank and business creditors. Be careful you don't become an unwitting assistant to an unscrupulous action.

2. Try to get a thorough understanding of the problem that you are asked to assist in solving. Are you an expert in the particular area? You should work only on those cases in which you have the knowledge and experience to successfully meet a challenge in court.

2

3. Do you want to undertake this particular assignment? In reviewing all the facts and participants in the case, you should take the time to determine whether it will be an asset or a liability to your business and reputation. Sometimes a poor choice of client, attorney, or case will have an adverse effect on you - even if your side wins.

4. Do you have the necessary time and capability to do the type of job required of the case? It is difficult to emphasize enough the importance of thoroughness in this type of problem solving. When a case goes to court, the expert must be thoroughly and meticulously prepared to assist his client. If you are not able to offer such, don't undertake the assignment.

5. Does a conflict of interest exist? Are you involved in a different case or problem in which the prospective client is on the other side? Are personal friends, acquaintances, or clients involved on the opposing side? Some discreet query into the matter will give you the necessary information on which to judge as to whether to work with this client.

These are just a few of the major areas of consideration that need to be given thought when asked for assistance. There are many others, of course, but these are considered mandatory to check out.

One last thought before we get into the various client types. Objectivity is a must in this business. It is very easy to become sympathetic to a client - especially if he has been grossly wronged by another party. Sometimes it makes one's emotions rise to the defense....even to the point of personally taking up their crusade. My advice: don't. That is the attorney's job. No matter how one personally feels about it, feelings and prejudicial thoughts should be kept under wraps – don't talk about it to anyone – except perhaps, yourself. The best way to serve a client is to do the utmost in presenting the facts so that the law can take its course. Remember, too, the client's opponent also will have experts to do battle with your opinions. As in all things, do the best job of which you are capable....the rest is up to your client's attorney.

With these thoughts in mind, let's explore who these clients are and get to know a little of the world in which they operate. Basically, they are categorized into the following groups:

1. Insurance Companies and Consulting Insurance Adjusters
2. Owners of residential and commercial properties
3. Property Managers, Real Estate firms, Bank Trust Departments
4. Businesses and Lessees of commercial property
5. Buyers of real estate: commercial and residential properties
6. Developers, Contractors, and Owner/Builders
7. Attorneys
8. Architects, Engineers, and other professionals
9. Governmental Agencies: Cities, Counties, State, Federal & Special Agencies
10. Courts

1. INSURANCE COMPANIES AND CONSULTING ADJUSTERS

INSURANCE COMPANIES. Contrary to what the agent may say, insurance companies are in business to make money. The person who lauds the benefits of owning a policy with his company is telling only half the story. When a policy holder has an occasion to file a claim under the terms of the policy, that claim is funneled through the system and eventually lands in the Claims Department. These departments are staffed by people known as Claims Adjusters. Adjusters determine if the claim has validity, whether it falls within the scope of that owner's particular policy, and the extent of the claim. They need to have complete information - in order to process the claim - including quantitative and qualitative descriptions that give a clear picture of the problem. From this, they will arrive at what they consider a fair market value for settlement. It is in this area of determining the cause of the problem, repairs if required, and value that the expert renders his assistance.

Most of the average person's experience with insurance companies is as a customer – a policy holder whose contact with the company was through sales personnel. These sales people are informative and try to impress upon the buyer the advantages to be gained by holding a policy with their company. Their job is to sell policies and, as such, are friendly, courteous people. When a claim is filed, the salesman refers them to the claims department where the adjustment team comes into play.

The adjuster's personality is quite different from the sales person. He does not have the problem of trying to please a policy-holder. Though he is usually friendly, his function is to process the claim, protect his firm and save the company money. He will examine the case from all aspects to determine the precise problem, the apparent cause, who is at fault, and most importantly, is the insurance company liable for payment under the terms of the policy. Consequently, the insurance adjuster is a knowledgeable, often tough, no nonsense type of person.

In working with adjusters, I've found them to be cordial people willing to work things out. However, the facts are their jobs depend upon their ability to correctly analyze the problem and, hopefully, to save the company money. The relationship that an expert has with an adjuster normally begins with a call for assistance. He is asked if he would be interested in assisting with a problem that is within his area of expertise. Adjusters request this assistance when conditions are outside their own knowledge and feel the need for backup from an expert - especially if there is a chance the case may go to court. The expert's relationship with the adjuster, then, becomes one of gathering all documented information pertaining to the case and often includes visiting the site or place under consideration for purposes of inspection and observation.

The adjuster will usually require an expert's written report that will state the problem, the cause, any related problems resulting therefrom, who or what was responsible, why it was caused, was there negligence involved from third parties, and were any laws or code violations noted. Sometimes, as in the case of damage, the adjuster will request calculations and plans for determining the cost of repairs. This is the role of the expert in his relationship with insurance companies. Large insurance companies usually have sizable claims departments employing many adjusters.

CONSULTING INSURANCE ADJUSTERS. These are private companies that serve as the claims department for those insurance companies who do not keep a staff of adjusters on board. Such insurance companies usually have a small claims section that processes the claims and then sends them out to a "consultant adjuster" for investigation and determination of the validity and extent of the claim.

These private adjusting companies are very similar to the in-house adjusting sections of the large insurance companies. The only difference being the livelihood of these private adjusters requires their maintaining a good relationship with their insurance company clients. Because of this, they are a little more inclined to deny claims and may have an attitude that the policy excludes more than it actually does. Because most policy holders know very little about their coverage, and even less about the legalese of the policy text, it is a simple matter to deny coverage unless a knowledgeable person is brought into the case, which is often an attorney.

The personalities of these adjusters are very similar to those of the insurance companies except that they are probably tougher and somewhat more argumentative than their insurance company contemporaries.

When the need arises, the private adjuster will make contact with the expert requesting much the same information as the company adjuster - problem cause, extent, code violations, etc., ending with a comprehensive report. Experts are often requested to provide additional data that will back up their findings - such as plans, models, cost estimates, etc., for the repairs, etc. Unless they also are experts at cost estimating, it is better to recommend - or hire - a qualified estimator or contractor to produce one in writing. The main point here is not to get caught in a situation where the expert is on the fringe of his knowledge. If the case goes to court, his "guestimate" may backfire as an opposing attorney will make mileage on his lack of knowledge. This would further open the possibility that other information supplied by the expert may also be challenged as unreliable thus reducing his credibility as an expert. Attorneys think like that.

The main point in working with both company and private adjusters is that they are seeking information that they are not able to gather themselves. And, they want this information from a recognized expert - one who is well qualified in his field to render an opinion that will stand up in court.

2. OWNERS OF RESIDENTIAL AND COMMERCIAL PROPERTIES

OWNERS OF RESIDENTIAL PROPERTY. As the name implies, these are people who live in and own their residences - as differentiated from those people who own residential property as an investment or business. Owners will call an expert when they have a particular problem. It will generally be something that has failed, or is questionable, and they need advice as to what to do about it. It may be some condition in their residence needing repair or concern about the quality of work that a contractor may have done for them on their properties.

They seek the expert's advice to identify the cause, the extent of the problem, and what will be required for repairs. Secondly, they will want to know if it may be insurance related in hopes of having coverage for reimbursement. Essentially, they are looking for answers to assist them out of their dilemma. Residential owners are, as a general rule, less knowledgeable and more emotional in their approach to problems than commercial property owners. This investigative work will culminate in a report to them stating the expert's opinions and recommendations.

OWNERS OF COMMERCIAL PROPERTY. These people are more business-minded than residential owners. They are the owners of what is generally called income property; this includes buildings from single family residences, duplexes, and multi-family residential complexes, to business offices, and retail and commercial buildings of all types. In contrast to residential owners, who often display more of an emotional involvement with buildings, the commercial owner usually is concerned with bottom line dollars and cents. Their need for an expert generally stems from problems arising in their properties requiring repair or for a poorly done contractor's repair job. Usually an investigation and analysis culminating in a report is required. Often, remedial plans for repairs are also requested.

There are some commercial property owners, when faced with repair problems that fall within the scope of their insurance policies, who will try to coach the expert into wording his report to cover items that are adjacent to, but not a part of, the problem - in order to pick up additional repair or maintenance work. The expert has to be frank in his replies to this kind of coercion. As an example, let's say that a roof member was damaged and in need of repair. The fix would be to do what is necessary in regards to repair or replacement of the member. Perhaps the roof membrane above the member would need some attention because of excessive deflection or sag. I have had cases where the owner would try to coerce me into editing my report to get an entire new roof out of the situation, though not warranted. Owners who try this usually own older buildings, have neglected their maintenance, and thus try to take advantage of an opportunity. Needless to say, the expert's credibility could be easily lost by succumbing to such an undertaking. Also, I have seen commercial property owners become highly emotional about problems with their buildings - especially if they don't appear to have a financial way out. Fortunately, most owners of commercial property are very pleasant business people who understand what is involved in maintaining and repairing their properties and act accordingly.

3. PROPERTY MANAGERS, REAL ESTATE FIRMS, BANK TRUST DEPARTMENTS

PROPERTY MANAGERS. Property managers are individuals or groups of people in firms who manage commercial and residential properties for absentee owners. The types of property are usually office and residential building complexes. These managers work with the building tenants for lease space. Leases vary from full service, which includes all services, utilities, taxes, maintenance, debt retirement, etc. that are included in the lease, to the type where the tenant assumes all obligations of the owner. The manager's job is to collect rents, lease space, handle tenant problems, maintain the buildings, landscaping, and all other duties for the upkeep and protection of the properties.

Property managers receive all complaints from tenants in the buildings under their control for any conditions that they find to be a problem in their particular area of lease. These range from leaking roofs to major structural problems. The property manager usually has someone who will perform necessary minor repairs when they arise. If problems develop that are out of the scope of the building maintenance personnel, the manager will then call in an expert to determine what needs to be done. As in cases with other groups of owners, the same information is requested: cause, repair, cost, time element, who's responsible, etc., and a verbal or written report - sometimes both - depending upon the situation.

Property Managers of residential property work primarily with those who live in the complexes; that is, the dwellers of the units. The manager's job in this area is difficult in that there are many types of personalities with an equal number of problems to handle in the course of their business. In so doing, they are firm and somewhat hard-nosed in their work. This attitude carries over into the relationship with an expert who may be called upon for assistance. They have a tough job, most handle it well and are usually courteous to professional people. It has been my experience that most of these managers have limited knowledge of construction.

Property managers of commercial properties, that is, office complexes, retail stores, and other business properties, are often more professional in their approach to problems. Perhaps it is because they work with business people who know and understand the ways of commercial enterprise. There is a considerable difference between working with the manager of a business and a person who is a dweller of a rental unit; this difference is reflected in their personal mannerisms and approach to the management concept of properties. When commercial property managers receive information as to problems from their tenants, they usually visit the building to make a firsthand appraisal of the situation and then call for assistance as needed. Often, the problem can be remedied by a competent contractor; if it is more serious, an expert – often an architect or engineer - will be called for analysis and recommendations for solution.

REAL ESTATE FIRMS. These firms have a somewhat broader scope than the property manager function, but they generally do the same work as the residential and commercial property manager people. Real estate personnel extend their boundaries into the sales and purchases of commercial and residential property from the standpoint of being brokers.

Real estate brokers usually become involved with an expert when there is a proposed sale or purchase. Many sophisticated investors put an inspection clause in their purchase contracts stipulating that the purchase is subject to a professional's report of the property's condition. The expert will then be requested to examine the building to determine its condition and list any potential problems in a comprehensive report.

Brokers are a little more cagey than most property managers. A good many property managers are brokers, although they don't function in that capacity unless a situation occurs where they can earn a commission from a sale. A number of people in property management also are investors or part of investor groups who look for opportunities of value that occur from time to time.

BANK TRUST DEPARTMENTS. A number of people who have wealth in the form of residential and commercial properties retain the services of banks to manage their properties - i.e. retired persons, widows, etc. For a fee, a bank will manage these properties and account for them in a prudent and, hopefully, profitable manner. Another area of wealth is property that is left in a trust as part of an estate, with management by the bank. This often includes large commercial and residential properties. Again, this is a property management situation wherein the bank trust department functions as such.

The bank officers who perform this management function are probably the top of the line of the property management field, possibly because of the image they have to portray as representatives of the bank. On a personal basis, they are much the same as the other two groups of property managers, but on a professional basis, they operate in a much more professional manner. All business transactions are conducted with paperwork to follow any verbal conversations.

Whenever these bank management officials have a problem, they respond very quickly to call in an expert for an analysis and report. The expert's relationship with trust officers is probably one of the most enjoyable relationships of all the managers. They are professional and do not operate on a gut level plane. They are personable, businesslike and, as bankers, they are courteous and thorough in their actions and communications.

4. BUSINESSES AND LESSEES OF COMMERCIAL PROPERTY

These are individuals and representatives of business firms who are lessees of commercial property. The lessor, in this case, can be either an owner or a property manager. When a problem persists without relief from the owner or manager, the lessee will call in an expert to analyze and document his findings of the problem so that they will have legal basis to pursue the responsible parties for correction. These people usually have little or no knowledge of construction and depend upon the expert to present his findings and opinions as to problem cause, responsibility, repair, cost, etc. If no relief is forthcoming from the property manager, legal action usually follows.

5. INVESTORS IN REAL ESTATE - COMMERCIAL & RESIDENTIAL

This group represents those who are buyers and sellers of real estate. They may be individuals, partnerships, investment groups or companies, but their main business is buying and selling commercial and residential real estate properties. These are knowledgeable people. They have a good grasp of construction in general and the ability to determine if the condition of their proposed investments is sound - or reasonably so. Their range of property involvement varies from shopping centers, apartment complexes and commercial buildings to high rise office buildings, hotels and retail/commercial complexes.

Prudent investors usually call in an expert to look over a proposed purchase to render a report on all findings. What they are seeking, primarily, is the discovery of some latent condition or problem that would require future expenditures for repair. Their reasons are evident: they want to avoid spending additional moneys for repairs that were unknown or undisclosed by the seller. There are laws on the books in some states that require disclosure of all known defects or problems with properties by the sellers; litigation often follows for non-disclosure of defective items. However, an expert's professional observation often discovers previously unknown defects. In some instances, the expert's report is the basis for a reduction in price or even cancellation of a purchase. Prudent buyers usually insert a clause in the purchase agreement that the sale is subject to a satisfactory report on the property by an expert.

6. DEVELOPERS, CONTRACTORS & OWNER/BUILDERS

Those who compose this group are essentially construction people well versed in the field. Personally, they are knowledgeable, direct, and often hard-nosed individuals who deal with problems on a daily basis and are fully capable of holding their own in most situations. Also included in this group are those who are adept at bluffing and bulldozing their way through difficulties in hopes of frightening off lesser would-be adversaries.

DEVELOPERS. Developers are usually individuals or firms that specialize in the development of new construction projects. New shopping centers, apartment/condominium complexes, office complexes, and the like make up the bulk of the work in this area. There are also developers who specialize in the rehabilitation of existing buildings in redevelopment and other older areas.

The developer usually works on sizable projects and either hires a general contractor or has one in-house. He arranges his own financing for his projects with banks and lending institutions and controls the project from start through completion. Developers use experts to assist in solving construction-related problems much like the general contractor. The information needed and presented by the expert forms the basis for future court action should the need arise.

THE GENERAL BUILDING CONTRACTOR. The general building contractor's main function is to build things for others. His expertise is executing the

15

plans and specifications produced by the design team and to erect the buildings and structures contained therein. He will call an expert when he has a problem that he or his staff cannot solve. Usually, it is a problem that requires a great deal of expertise because most large contractors have technically capable people on staff who handle the multitude of daily problems encountered.

The small contractor, because of his size, does not have the resources of the larger ones and will call an expert more often. Small contractors usually are craftsmen who have started their own business and continue to operate in their trade; they usually don't have a staff to draw upon for assistance. Coincidentally, it is usually the small contractor that is involved in a large proportion of the problems as he usually tries to solve them himself - without technical assistance - and often makes conditions worse. A large number of problems that end up in court are brought about by small contractors who get in over their heads.

The general contractor is the party who contracts with an owner and is responsible for the construction or repair of a project. He is the captain of a team composed of assistant contractors who aid him in the completion of the work. Thus the name: "general" contractor. He supervises all phases of the work, including scheduling and coordination of his assistants who are known as "sub-contractors." These are specialists in their respective trades and include carpenters, concrete workers, plumbers, electricians, roofers, masons, earthwork and paving workers, and the like. Each sub-contractor cooperates with the others to complete the project. It is during this construction process that disputes may arise between the general contractor and his subs or between sub-contractors. Often experts representing each side are engaged to investigate a problem and render their findings, which may be used to settle the matter or be presented later during court litigation. The problems, which can take on a variety of forms, are usually mistakes that were made during the construction process.

OWNER/BUILDERS. These are usually small investor/owners who act as the general contractor and hire subs directly rather than employing a general contractor with responsibility for the project. Owner/Builders operate primarily as people who build projects that they will continue to own and occupy after completion. They usually arrange their own financing, have their own project design concepts, and hire their own design team. This is usually the embryo state of becoming a developer. These people will occasionally require an expert when they have a problem with one or more of their sub-contractors who have not performed their tasks according to contract. The expert's job is to compare the work as completed with the contract plans and specifications and to author a written opinion of the affected work at that time, which will delineate discrepancies found. The report then becomes a legal instrument usable in court.

7. ATTORNEYS

Attorneys use experts much like general contractors use sub-contractors. The attorney, like a general contractor, cannot do the job alone; he needs those with expert knowledge to assist him by supplying information pertinent to the issue in question. He is similar to the quarterback on a football team; he has an overall plan to attain his goal and needs each of his players to know their task to accomplish this. His experts, through their investigations, study and analysis, supply him with information to prepare his case, whether it be for the plaintiff or defendant.

Law firms vary in size from the sole proprietor office to large partnerships or corporations composed of many attorneys. As a group, those in real estate and construction litigation require extensive use of construction experts and are probably the best source of clients in that field.

8. ARCHITECTS, ENGINEERS AND OTHER PROFESSIONALS

Often there are situations where architects, engineers, or other professionals have a problem in their practices or one in which they are called as an expert. If they have a claim against their practice, they will usually require assistance from experts in their field with similar knowledge. If called as an expert, they may require assistance from another professional in a related field for specialized information. For example, a structural engineer investigating a building foundation settlement problem may request a geotechnical engineer to provide information on soil characteristics, such as classification, bearing strength, expansive characteristics, and related information.

If a professional person or craftsman has a legal action filed against him, he may require the assistance of an expert to conduct the investigation for the case and work with his attorney. This assistance also extends to experts of the same discipline to render a second opinion; specifically, someone who is not emotionally or economically involved.

9. GOVERNMENTAL AGENCIES: CITIES, COUNTIES, STATE, FEDERAL, AND SPECIAL AGENCIES

From time to time, various governmental agencies have problems in which they have to reach outside their own technical staffs to obtain the opinions of other experts in the field. As an example, if there were a major deficiency discovered in or during the construction of one of their buildings or some similar situation in which they were directly involved, they would normally call an expert in that field for assistance. His task would be to conduct an investigation of the problem culminating with his written opinions and recommendations; as such, it provides documentation to be used for possible future litigation.

Relationships with governmental agencies usually require meeting with the department heads responsible to the political process for their particular agencies. Such meetings outline the scope of the problem, the available data and any contractual arrangements necessary to conduct the investigation. Normally, the expert for this type of assignment needs a well-established background, professional credibility and a professional reputation. The more prominent people in the forensic field are usually contacted by this type of client.

10. COURTS OF LAW

From time to time, during the trial process, opposing sides come to loggerheads as the result of expert opinions. Often, when such occurs, the judge will appoint an independent expert to perform a study and offer his opinion to settle the matter in one direction or the other. The client in such cases is the court with the judge acting as the responsible party for contractual arrangements. The duties of the expert are much the same as in other investigative matters concluding with his opinions and recommendations. The expert selected for this assignment is required to be acceptable to each side of the litigation.

As can be seen, there is a wide range of clients to be served in the expert witness field of endeavor. All have a variety of reasons for desiring the information the expert can produce, but they all follow the same general pattern. Essentially, they want to know why a given incident or problem has occurred; what was the cause, who is potentially responsible for it, what can be done about it and, if applicable, how much will it cost to employ the remedy.

The problems the expert encounters are as varied as the elements that comprise the various fields of industry. As in the construction field, they may be the design plans and specifications, the materials used, the manner in which the materials were installed, the labor relationships between the various crafts and trades in the process, contractual performance, monetary disputes, and so on. The problems range from the very simple - as a defect in a dwelling - to the very complex - as in a structural roof failure in a covered sports stadium. The expert has to be knowledgeable in his field of expertise. He should have formal knowledge of his craft as well as extensive field experience.

CHAPTER 2: THE EXPERT WITNESS

Webster's Dictionary defines the word "expert" as "one who has acquired special skill in or knowledge of a particular subject; having, involving or displaying special skill or knowledge derived from training or experience." It continues on to define the word "witness" as "one that gives evidence; one who testifies in a cause or before a judicial tribunal." The term "expert witness" has been defined in numerous legal writings as: "any person who has the special knowledge, skill, experience, training or education necessary to become an expert in a field may be qualified to testify as an expert."

So the person who is to serve as an expert witness must prove upon request that he possesses the necessary abilities. Attorneys are usually very careful to query potential experts as to their qualifications knowing that they must successfully pass a challenge by the opposition to have their testimony admitted in court. Anyone serving as an expert may, at any time during the process from deposition to the courtroom, be challenged for such proof.

To illustrate this point, I was retained as an expert in a case involving wood frame construction along with another structural engineer. While he was on the stand, the opposing attorney challenged his expertise in the area of lumber grading. Structural engineers are not lumber graders, but they do have sufficient knowledge to determine the general characteristics between grades of structural lumber, which was all that was needed in this particular instance. The engineer, accurately knowing the difference between a lumber grader's knowledge and that of a structural engineer, admitted that he was not an expert lumber grader. The attorney immediately requested the judge to dismiss the witness because he, by his own admission, declared he was not an expert. The judge dismissed him from

21

testifying. This act completely threw out not only any testimony the engineer had given regarding lumber, but all other information of a structural engineering nature as well. Whether this was just in this case is certainly debatable, but the end result was the engineer's total testimony was lost.

When my turn came to be cross-examined, I was also asked if I was an expert in lumber grading. After observing the first engineer's experience, I answered in the affirmative and would submit to his questions to prove such. The attorney did not know very much about lumber, so the questions he asked would have been easily fielded by the first engineer. Withstanding the challenge, I continued on with my testimony. The point is the disqualified engineer was thinking on a broader scope than the attorney and, because of this misconception, was disqualified.

If you know your basics, stand your ground and put them to the test. Another point to be made is that attorneys, by their position as officers of the court, usually try to give the impression they know everything about that which they are discussing; most do not and are bluffing. An adage in the law is "never to ask questions of a witness when you don't know the answer"; so their queries are usually fundamental. The point to keep in mind is that you are the one who has the expert knowledge, not the attorney, that's why you are there.

PROFESSIONAL QUALIFICATIONS. As previously stated, the client, be he an attorney or other party, wants to be assured that you have the necessary expertise to assist him in finding a solution to his problem. In order to provide such information in an organized manner, thus saving time and effort as well as adding a professional touch, a resume of your qualifications should be presented to the client. The legal definition for a document containing this information is know as a Curriculum Vitae or "C.V." The following outlines those important items to be included:

1. ACADEMIC BACKGROUND. This is a chronological accounting of your formal education including colleges and universities attended, degrees earned, scholastic accomplishments, special areas of study and expertise, and all scholastic awards earned, as well as their dates.

   If your skill, knowledge and expertise were gained in the field of practice, as is the case of many technicians, contractors or other craft disciplines, list all correspondence courses, company programs, apprenticeship training, special classes and seminars in which you learned your craft. List also all certificates and awards received, papers written and/or ideas that were adopted by employers, including dates, received.

2. PROFESSIONAL ORGANIZATIONS. Present a list of the organizations to which you belong that represent your field of expertise. Examples of such organizations in the construction field would be: the American Institute of Architects, American Societies of Civil Engineers, Electrical Engineers, Mechanical Engineers, Structural Engineers Association, Associated General Contractors of America, The International Council of Building Officials, The American Institute of Steel Construction, The American Concrete Institute, and the like. Obviously, a similar listing would be for other fields of practice; i.e., accounting, medicine, sales, etc. Give your affiliation with the organizations, offices held, length of membership, and any pertinent information regarding your involvement.

3. PROFESSIONAL LICENSES. State the professional licenses acquired, the dates obtained and the States of licensing. Examples are: the Architect's Professional Registration, Civil, Electrical, Mechanical, Structural, Geotechnical Engineering Registrations, Engineering or

General Contractor's License, Registered Construction Inspector's License or any license granted by the State for specialty contracting, such as: roofing, plumbing, electrical, painting, etc. If you have been granted qualification certificates as a special inspector or certified technician in a particular trade, list these also. The same applies, of course, to other fields of practice.

4. WORK HISTORY. This begins when you first entered your field of work. For many, it will be after graduation from college; for others, it may be their first job in the industry. What you are establishing is a chronological history of your work which, using your academic skills, developed your knowledge and experience to the present.

While it is important to include detail to show that you do have experience in a particular area of work, do not list every employer if the work was of a short period and the tasks were the same as other jobs of longer duration. The important thing is to show that you have had sufficient experience in the particular areas of your expertise to qualify as an expert. List prior legal cases where you have served as an expert, if any, by category and dates. Don't overlook the value you may have gained from military experience as related to your field giving your rank and responsibilities. It is very important that you do not stray into areas either on the fringe of or just outside of your particular expertise. If projects are undertaken in these areas, you will be very weak in your assistance and will do both your client and yourself an injustice. Stay in the areas where your strengths are the greatest.

When preparing the work history section of your resume, place the most emphasis on your experience in

24

the area in which you serve your client. List previous service as an expert in that area of expertise, if any. If the problem in which you are asked to address is in the field of wood construction, for example, place the emphasis on your experience in that area. It will not do much for your credibility if the case is about wood and you extol your experiences in concrete construction. To place emphasis on your experience, include a list of projects on which you have worked. The list should be broken down into types of buildings, structures, etc. and how many of each during the course of your career. It also is impressive if you can give a gross dollar amount of construction costs represented by your list. Again, be sure you have the expertise to serve your client in the area of his need, avoid fringe-of-knowledge areas.

5. PROFESSIONAL CONTRIBUTIONS. List all items that you have presented to your field of endeavor. These may take the form of papers written, books, ideas that have been developed, practices that have been adopted, inventions patented and the like. Include all teaching positions in schools, colleges, universities, special seminars, and night education classes in your field. These may be as a part-time or full-time instructor or on occasion as requested. Also, list all awards and certificates of merit received for your contributions to your field of expertise.

6. EXTRA-CURRICULAR INFORMATION. Add to your resume all areas of public and civic participation and accomplishments. These include offices held in civic organizations, government, politics, as well as private service organizations that work for the betterment of your community. If you have some unique hobbies in which you are actively engaged, they should be listed also. If you are a pilot and fly your own plane, a

published writer, a well-known lecturer or of some other note-worthy endeavor, such should be added to your list of accomplishments.

7. BROCHURE. Since it is very difficult to prepare a specialized resume for each client, a good solution is to prepare a brochure of all information that is of a more permanent nature. This would include all items that would remain unchanged throughout time and would give a general overview of your background and the services that you offer. The brochure would then be sent along with a brief resume listing those pertinent items that slant toward the client's needs. I revise my brochure about every three years, depending upon the timeliness of the information contained therein. My resume is revised for each request. Both are then sent, including a fee schedule, to the client for his information and file. If you have received letters from former clients in appreciation of your services, include copies to add validity to your performances as an expert. You also may include past clients as references providing you have their permission.

The resume and brochure are valuable in depositions and court proceedings. They save time in qualifying the expert as most of the information desired is already available in written form. Without this information, the questioning process can go on for a lengthy period of time, depending upon the attorney. These queries place significant demands upon the memory resulting in mental fatigue and, more often than not, some vital piece of information is omitted that would enhance your qualifications. Armed with the resume and brochure, the opposition normally will ask only a few questions on the information presented and possibly some elaboration on your background and expertise.

The Curriculum Vitae (C.V.) resume serves one vital purpose. It presents the expert's qualifications in written form giving all information needed to make a judgement on his expertise. It literally stands as the expert and pays dividends to make it brief, viable, yet as complete a picture, as possible.

PERSONAL QUALIFICATIONS. While the resume and brochure will present the best posture for professional qualifications, there still remains the impression that the expert will make in person. This impression will be made upon a number of people: (a) the client with whom he will be working; (b) the client's attorney, if not the same person, and his experts; (c) the attorneys for the opposition and their experts; (d) the judge of the court; and finally, (e) the members of the jury and court attendees. So as you can see, it is important to make the correct impression personally. The following are some areas of consideration for checking personal qualifications:

1. DRESS AND GROOMING. Because the expert is, in reality, a performer, he should be sure he is properly groomed for the part. His clothes should reflect the professionalism he is to represent. In meetings with the various individuals and groups involved, grooming should consist of attention to presenting a clean, tidy appearance with fresh clothes, shined shoes, appropriate haircut and manicure. The dress code varies with the tone of the meetings; the following has been found to work to the benefit of the expert during encounters with others:

    A. CLIENTS, ATTORNEYS AND OTHER PROFESSIONALS. The expert is judged by his outward appearance at each meeting. As such, he should present a prudent and conservative image reflective of his thoughtful approach to the task.

Nonflamboyant suits, tasteful sports clothes, shirts and ties in generally subdued colors are usually an acceptable mode for dress. Loud colors and mod-style clothes give rise to questioning the seriousness with which he will approach his job.

B. DEPOSITIONS, ARBITRATIONS AND MEDIATIONS. In meetings with attorneys, the dress code can be either as illustrated above or in more conservative suits, shirts and ties. In depositions, the opposition is making judgements as to the expert's general presentation of appearance as well as his mental capabilities. They are literally sizing him up as to the impression he would make upon a jury. Arbitrations and mediations are slightly less formal meetings than in court, however, it is prudent to dress conservatively for the occasion.

C. IN THE COURTROOM. The credibility of the expert is on the line in the courtroom. His attire should be a reflection of his position and information he is to present. He should wear a well-tailored, conservative suit of medium to dark blue or gray, white shirt and a conservative tie. No loud lapel pins or personal jewelry, such as chains or bracelets, etc. should be worn. Conservative lapel pins such as those reflecting a professional organization are favorable subtle wrist watches and wedding rings are a plus. Flashy ornamentation detracts from the conservative appearance and can become distractive to jurors while listening to his testimony. Notice how the attorneys are dressed and groomed; they usually represent the norm for courtroom appearance. That famous line from the 1960 presidential political campaign applies equally to this situation: "Would you buy a used car from this man?"

There are times and places, of course, when the dress code should be learned prior to making an appearance. As an example, it would not be very helpful to an expert to be dressed in a conservative business suit while testifying before a jury in court where the general dress was western clothes, as is the case in some areas of our country.

D. SITE INSPECTIONS. The dress code for going into the field for inspections would be to reflect the general type of clothing as would be found by those working in the area. In some locations, a person in a business suit could meet with a reluctant reception merely because of his attire. If he had been more modestly dressed, the resistance may have been less. Conversely, showing up in work clothes in a place of business where suits are the code, the expert may be accorded less respect and cooperation than would be expected for someone of his profession. Again, dress for the occasion. Check on these items prior to the inspection.

2. SPEECH. The resume and brochure give an overview of the skills, knowledge and experience of the expert. The personal presentation he makes with his appearance should fit the mold that has been created. All of this preparation, however, can be lost when he opens his mouth if he doesn't know how to speak.

Most people who have reached the stage of their professional life where they can qualify to be an expert have learned how to present themselves orally. However, because it is such an important part of the expert's tool kit, some guidelines are set down for consideration:

A. TIMING. It is of the utmost importance to know when to talk and when to keep quiet. It is a well-known fact that those who say little are accorded as knowing more than they do – until they open their mouths. I have found the best form of talking that the expert can engage in is that of asking questions. This applies to conversations with a client, an attorney, and during the information gathering of the investigation. Some elaboration of your personal qualifications and personality is in order in working with a client and other experts, but generally, the less said the better.

During the deposition process, it is very important to respond to questioning with only enough information to provide an answer. Such responses should be short, succinct, to the point, and then terminated. More will be presented on this phase of the expert's role in the chapter on depositions. The oral testimony of the expert in an arbitration, mediation and the courtroom will be covered in later chapters.

B. CONVERSATIONS. In presenting oneself to others, the expert should keep in mind that there are always two sides to one's personality. One is the personal side as when conversing with friends and acquaintances, the other is professional as when engaged in business practices. In working with clients, as well as others representing your side of the case, the expert should present his professional side allowing himself to become personal only when the occasion would warrant. While he should strive to become business friends with those whom he works, he should avoid becoming too personal. Keep to the job at hand allowing only for necessary amenities.

When meeting with the members of the opposition, the expert should always be cordial, but not enter into any friendships. He should be professional and impartial with everyone connected to the case. While his job is to assist his client to the best of his capability, he must remain objective in his approach to the tasks involved. The slightest hint of bias on his part could easily cause the opposition to attempt to nullify his participation in the case. After all, the purpose of procuring an expert is to obtain the facts and his unbiased opinions on the issue. It is quite possible to find that the expert for the opposition happens to be one of your professional friends. In this case, the best course is to acknowledge his presence in the same manner as any other person. Do not, however, engage in any familiarity or small talk as most friends do when they meet. Especially, do not talk about the case at hand.

C. LANGUAGE. In conversations with clients and the opposition, keep to commonly used words when describing the technical items of your field; avoid the vernacular of the trade or profession that is not fully understood by lay persons. It is important to be clearly understood in both your written and oral use of language. Misconceptions can easily arise when the spoken word on a given subject is not fully understood by those listening to your comments and may require an unwanted and unnecessary explanation at some later date.

3. ATTITUDE. The expert's attitude should reflect a willingness to help, demonstrate professional objectivity and be a seeker of truth. There are occasions during the investigative process where the truth cannot readily be determined, that is, the real reason why the event has occurred. Sometimes it can be obscured because of statements made by the client or others. The professional attitude is to seek the true cause of the problem, wherever its origination, and record the facts as they come to light.

When confronted by the opposition, always maintain a cordial attitude although not a helpful one. The expert should be autonomous in his position going about his tasks and reporting his findings only to those authorized to receive the information. Often when investigating a problem wherein both sides of the issue are present, adverse experts will make queries and ask questions to try to determine your position or to undermine your obvious approach to the issue. The best way to nullify such attacks is to politely ignore their volleys stating, if necessary to avoid confrontation, that they will have an opportunity to know your opinions later during depositions or trial.

4. PUNCTUALITY. It should go without saying that the professional approach to one's work is to do it with dispatch. Clients are always in a hurry to obtain information and one of the functions of the expert is to deliver such in a timely manner. This means that once the job is accepted, the process should begin as soon as possible starting with a schedule given to the client. If difficulties are encountered requiring additional time, he should be informed and assistance requested if it is within his power. Punctuality also includes arriving on

time to meetings, etc. with the appropriate information in your briefcase.

5.   KNOWLEDGE OF YOUR FIELD. To function as an expert, you should be well versed in your field of expertise. However, the technical advances of materials, methods of construction, computer-aided analysis and calculations, codes and regulations require an on-going schedule for keeping abreast of developments. Codes change with new additions every few years, as do manuals and client requirements. The best way to keep abreast of the latest developments is to keep files on the various categories of information pertinent to your field. By so doing, you will have stored information for your research when a specific need arises. It is impossible to keep knowledge of all things related to your field in your mind. The best place for storage is where you can find the needed information easily. A library of reference books, periodicals, papers, codes and computer data is of great assistance to the research phase of the work. Additionally, past cases on which you were an expert aid greatly when similar challenges arise.

With respect to codes, investigations of existing buildings or structures should follow code requirements for the time of their construction. Because codes change with each new edition, current requirements may be substantially different from previous editions and do not apply unless additional construction was performed under the then current code.

ADVERSE EXPERT WITNESSES. In cases that are expected to go to trial, arbitration or mediation, attorneys representing opposing parties will be seeking assistance from experts with technical knowledge pertinent to the issue in dispute. The success they have in procuring competent and qualified help really depends upon their own knowledge and ability to screen the competent from the incompetent.

Inexperienced attorneys seem to be more impressed with academic credentials than practical knowledge and experience. There are many people in the "expert" field who tout their services because they possess a Ph.D. degree in a field similar to the one at issue. College professors who spend their entire careers teaching the same courses to students again and again are well versed in theory but, as anyone who has been voyaging on the sea of experience knows, theory and practice are two very distinct and different fields. I have been on the opposing side of such professors many times and found a good percentage of them to be of the Ivory Tower-type personality - with an ego to match. Very few have had actual experience in putting their well-entrenched theories into practice – as in actually engaging in construction practice. I have no argument with the theoretical folks, they are an important and necessary part of our society. But some, usually the one's who like to testify, do not have the depth of experience that is gained from practicing in the market place where decisions result in large monetary expenditures and problem solving is a way of life. There is a big difference between grading student papers and signing one's name and affixing his seal to a multi-million dollar construction project.

So many times you will find that the experts representing the opposition will be quite well educated but do not have the necessary combination of education and experience that you may possess. The most effective expert is the one who is on his home ground with the task before him. Conversely, if you find that you lack the necessary qualifications because of a weakness in a given area, you may choose to bring in a consultant for assistance or else decline the opportunity and let it pass. In any event, opposing experts have a task in serving their client. If they are assisting the plaintiff, their findings and opinions must stand up under rigorous examination by their contemporaries on the side of the defendant. On the other hand, if they are representing the defendant, then their task will be to check the plaintiff's expert's findings for inaccuracies and unfounded opinions. In other words, their goal is to educate their attorney client in the technical aspects of the action while also searching for information that he can use to discredit the opposing expert's findings and opinions by negating his competence in the action.

Whenever I am asked to assist as an expert, one of my first questions to ask is who are the experts for the opposing side. The answer to the question gives me insight in sizing up the opposition, their probable knowledge in the field and their reputation. It is interesting to note that some of the most popular and respected names in the construction field turn out to have less knowledge than other lesser-known practitioners in the field.

One of the most important functions an expert can perform is in his ability to read the opposition's experts. As an example, I was serving on a case in which the opposition utilized two college professors as their experts. As one can imagine, they both gave long dissertations on the engineering principals involved in the case to prove their points. As expected, their explanations were highly theoretical and, in my opinion, did not represent the actual condition of the materials under discussion in the case. Investigation into their credentials found only one was licensed to practice civil engineering, the other claimed to be a mechanical engineer, but did not possess licensing registration. Needless to say, the latter's claim was illegal and it was against the law to use the title without State registration. My attorney client made good use of this information that he otherwise would not have known without a thorough check on the opposition's experts. Another clue to the opposition's experts is to determine, in the case of professional registration, the date of their licensing. In one case, in my past experience, was a report authored by a registered engineer. It contained many statements emphasizing his opinions that claimed his client

35

was without fault and spread the blame on everyone else who was party to the lawsuit. This barrage of accusations piqued my curiosity as to his qualifications and experience. My query into the time of his registration found it to be less than a year old. The conclusion drawn indicated that he was an enthusiastic neophyte who was not thorough in his investigation. His charges were found groundless during the succeeding arbitration.

With respect to thoroughness, it has been my experience to find that most experts who served on cases in which I have been involved, do not do their homework. They are satisfied with their first answer to a problem and then establish their position on that base. As most know from experience, the obvious answer may not be the true one. Only by continuing to thoroughly investigate all possibilities can the true reason be found and relied on with confidence. I have had opposing experts take offense and try to justify their position after listening to my reasons for the cause of a particular problem, obviously, they were embarrassed upon discovering their omissions and tried to save face by attacking my findings. In one particular case, I simply told the opposing expert, because he was also an engineer, to perform the calculations on his own if he wanted proof. If he had done his homework, he would have found the same thing as I thus saving both himself and his client embarrassment. You have to be thoroughly prepared to win battles.

As in any confrontation, be it an action in which you are personally involved or one in which you have been asked to participate, it is to your advantage to be sure you are properly qualified professionally. You must do your homework, prepare your plan of action, present yourself properly, express your thoughts clearly and intelligently – and know your opposition thoroughly.

CHAPTER 3: THE INVESTIGATION

The investigation the expert conducts for his client is at the heart of the case. From it will be collected all available information that will form the basis for his analysis, opinions, conclusions, and resulting recommendations. From the very onset, this is a meticulous and methodical search for all relevant material. Beginning with the client interview, it continues on to collect information relating to the issue for compilation into special reference categories to be used during the analysis phase of the work. Items to be collected include documents such as records, correspondence data, plans, calculations, depositions of opposing parties, etc., as well as visiting the site for investigation purposes. It may entail the application of special tools or equipment in gathering technical data, the scheduling and interviewing of witnesses, documentation of information, and on to performing analyses and calculations as necessary. The resulting conclusions from this study form the opinions of the expert and express his recommendations. At the completion of the investigative phase, all information is placed in a comprehensive written report. As stated previously, the field of the author is construction. The following information pertains to investigative procedures and equipment related to this field. For those in other nonrelated fields, the general principles for the investigative work are essentially the same though materials and equipment may differ.

Interviewing the client. Clients have different approaches to presenting problems and discussing services. Some will want the expert to meet them at the site immediately – especially if it is a developing problem; others desire to discuss the case and fees before the start of any work. Insurance adjusters, who handle problems on a daily basis, will ask if the expert is willing to accept the assignment and, if so, will give all relevant information over the phone ending with a request for a written report. Almost all are in a hurry to get answers. There are some clients, usually unsophisticated building owners, who want the expert to perform "just a quick check" on a problem as they want to keep fees to a minimum. The expert should be careful in this type of situation. A quick look may not reveal enough information for the expert to form a studied opinion - and he may find himself in court defending his statement and wishing he had taken more time to get all the facts. Problems have a way of expanding beyond the original scope and the expert should press on until he is satisfied that all necessary items have been taken into account.

Regardless of how the approach is made, there are certain items to be completed before beginning. As stated earlier, a screening of a new client is a must. The first item of importance is to get a thorough understanding of his perception of the problem. It may require some finesse to gain this information as some clients may not be familiar with the technical aspects involved and view the whole thing with alarm. The prospect of an unexpected and substantial cost is usually the major worry associated with most owners – especially if its not covered by insurance. Secondly, the following series of questions need to be answered: (1) is the required scope of work within the expert's expertise; (2) does he want to undertake this particular task; (3) could there be a possible conflict of interest with other parties involved in the case; and (4) does the urgency of the situation require immediate action or is there enough time to organize for a thorough investigation. Time is usually of the essence and delay can result in a loss of valuable information as evidence may disappear rapidly as memories become confused and fade. These are judgment factors that need to be analyzed quickly during the initial contact. Answers to these initial queries are usually sufficient for a gut feeling of whether to continue the process or to pass.

Once the initial questions are satisfied, a closer look should be taken at the client and his position in the matter. Is he a plaintiff or defendant? Does he desire the information for his own use or does he represent a third party? Finally, the question of services and fees should be discussed. Some clients have no idea of the fees charged by experts, it is prudent to tell them the range up front so there are no misunderstandings later after the work is well underway. Even if the initial contact is over the telephone, firming up proposed services to be rendered, along with approximate costs, should be discussed during the contact. Following this beginning, a personal meeting with the client is in order to complete the initial survey.

Based upon the assumption that the personal interview and the service/fee agreement are satisfactory, the most important item in the case must now be determined. This is to precisely define the problem, the extent, and any related problems deriving from it. Clients are usually full of ideas as to how things happened and reasons why. Experienced attorneys and insurance adjusters are quite adept at zeroing in on the probable cause of problems. The expert should listen carefully to what is said, ask questions for elaboration, and take notes on the information offered. He also should try to determine exactly what the client wants to know from the investigation and his reasons for desiring this information. This knowledge sharpens the directional focus and thinking for the expert. Sometimes the answers are obvious, at others some digging may be required. It is vitally important, however, that he understands his goal and works toward that end, few things can be more demoralizing - not to mention reputation damaging - than to solve the wrong problem.

At this point, it is well to be reminded that some clients may try to point the expert in the direction of the results they would like to see - which may differ with what the facts reveal. It is very important for the client to understand at the outset that the investigation and his report will be representative of the facts discovered and the expert's unbiased opinions in the matter. This is the area where some experts acquire the name of "hired guns" as they try to make the facts fit the outcome desired by the client. These people soon gain a reputation as such and are usually sought out by the unscrupulous. The honest expert who is methodical and thorough in his work will not have a problem testifying against these types of "experts."

The next item to obtain from the client is all documented information in the way of recorded data as plans, specifications, previous reports by other experts, depositions, and other written data that is connected with the problem. If items are missing that he deems necessary or desirable, he should ask the client for assistance in obtaining them. Clients, such as owners and lawyers, often have better luck in obtaining information from reluctant sources than do experts. Finally, he should obtain a list of all persons who may be involved with the case including names, telephone numbers, addresses, relationship, area of involvement, and the like. This also includes workers in the immediate area, eye-witnesses to the situation at the time of the occurrence, and people with an awareness of the problem before it became common knowledge. He should carefully identify all parties, their attorneys, experts, etc., that may be involved in the case. Once he has obtained all the information the client has to offer, he should remind the client to call his office if he happens to think of anything else he failed to disclose or any new information that has bearing on the case.

On returning to his office, the expert creates a case file for categorizing and storing each of the various types of information that will be gathered in the forthcoming investigation - beginning with the facts obtained from the client interview. The next step is for the expert to study all data for familiarity and, if necessary, make preliminary notes and necessary computations to assist in the forthcoming site investigation. After the information has been reviewed and understood, an itemized itinerary of tasks is then compiled homing in on the specific information to be collected during the site visit. A very important item to remember at this point is that the expert may have only one opportunity to visit the site as further access may be denied by the opposing owners or legal consul. It is paramount, therefore, for the expert to be as thorough as possible in constructing the itinerary and to follow it meticulously during the site visit taking all the time allowed.

Elaborating on the itinerary, it is a detailed list targeting those items that will reveal specific information needed to understand the causes of the problem. In some cases, they are readily apparent; however, more often than not they may be hidden and need digging to discover. The itinerary keeps the expert on track with its planned program. It should also contain the names of people at the site who are familiar with the problem – witnesses available for interview.

There are many times when items important to an investigation are not just lying around to be discovered. Often, as in the case of construction-related studies, portions of buildings and other structures require opening for access, as in walls, ceilings, roofs, etc. This procedure is commonly called intrusive investigation. The pre-inspection preliminary study will usually dictate if intrusive investigation is required. One should always be prepared to do what is necessary when visiting a problem site including obtaining permission for intrusive work if deemed necessary; as mentioned above, often there isn't an opportunity afforded for a second look.

Depending upon the type of investigation to be made, various types of tools are used depending upon the situation. In my investigative work in construction, some of the more helpful items used include:

41

1. A lined notebook mounted on a clipboard. Several pencils, a pen, a triangle straight edge, a six-inch architect's scale, a small magnetic compass and a thermometer.

2. Hand tape recorder. I find this device to be exceptionally useful. The recorder is a small, hand-held device that allows freedom of movement allowing more inspection time as it eliminates cumbersome and time-consuming note-writing. During an inspection, I record a running commentary on the conditions seen with appropriate opinion comments as applicable. The tapes are transcribed into computer and paper files immediately upon return to the office. It is also wise to carry extra tapes as well as batteries in the equipment bag, it is demoralizing to be recording or photographing and have the batteries go down. Unfortunately, it happens at the worst of times – when a second visit is out of the question. The notebook is used for special items as in the drawing of sketches, etc.

3. Cameras are standard equipment for investigations. I have four I use in my work. Two are 35mm, one is digital, and the fourth is a Polaroid. I shoot one camera with black and white film, the second and third with color, and the Polaroid for those photographs that are needed immediately; all but the Polaroid have date functions that mark the photos – they are indispensable for recall. The first camera is of standard size with a telephoto lens. The second is a small pocket size, also with a telephoto lens, that can be easily used in tight spaces when maneuverability is difficult. The third is a digital used for recording color images to be entered into my computer for color presentations to be included in the forthcoming report. All cameras are flash-equipped. I use good quality film with 12 & 24 exposures per roll,

depending upon the situation, and have the completed rolls developed immediately upon returning to the office. Once developed, they are inspected, identified and edited for inclusion into the file for future reference. Photographs are indispensable for both analysis and court documentation. Also, they have saved many a return trip to the site for lost or forgotten information. Carry extra camera batteries, too.

In the investigations I've performed, I find the regular use of video recorders to be less desirable than cameras. In my opinion, they don't work as well as cameras, especially in confined areas. The final review of the developed tapes is exceedingly slow and tiresome because of the preponderance of nonessential information. Editing certainly helps, but the audio portion distracts from viewing attention because of the constant dialogue interruptions. There are certain situations, however, where they are a good backup as additional information in combination with camera photos. Field practice sees very little video presentation at conferences, usually photographs are the norm.

4. I have three flashlights that serve adequately for inspection purposes. The first is a six-volt, spot flashlight capable of providing intense illumination of an object up to sixty feet away. The second is a three-battery standard spot/flood flashlight for general use. The third is a small, vest pocket, spot/flood flashlight used for tight spaces and carrying convenience. All lights serve their purposes very well and are indispensable for inspection inside most buildings and structures. Extra batteries and bulbs are a must.

5. I carry two ladders in my vehicle. One is a wooden twelve-foot fold-up ladder that packages down to a four

foot length, the second is a six foot aluminum step ladder. With these two, I can easily access most elevated locations. Almost every investigation requires some sort of climbing. Very seldom is a ladder of any type available at an inspection site. An additional thought pertaining to inspections is experts are often required to access into difficult areas often requiring personal strength, as such I've found that physical conditioning is a necessity for safe work.

6. Carpenter tools and other special items serve well in gaining access to hidden information and offer personal protection at the inspection site. As an illustration, I have the following group of items in my office to use as necessary:

> 1 - Water manometer for measuring floor and ceiling elevations
> 1 - 30 foot measuring tape
> 1 - 100 foot measuring tape
> 1 - 16 ounce claw hammer
> 1 - 4 foot wrecking bar
> 1 - 1 foot wrecking bar
> 1 - Cats paw nail puller
> 1 - 1 foot pry bar/nail puller
> 1 - 16 lb. sledgehammer
> 1 - 2 lb. sledgehammer
> 1 - Pair of bolt cutters
> 1 - Set of wood chisels
> 1 - Keyhole saw
> 1 - Tool set: wrenches, screw drivers, pliers, wire cutters, etc.
> 1 - 6 foot spirit level
> 1 - 4 foot spirit level
> 1 - 2 foot spirit level
> 1 - String spirit level

1 - 300 foot roll of string line
1 - Chalk line
1 - Engineer's transit, tripod & rod
1 - Hard hat
1 - Set of jumpsuit overalls
1 - Pair of field boots
2 - Pair of gloves
1 - Spider brush
1 – 4 x 6 foot plastic tarp

7. Other miscellaneous items include a pair of good quality binoculars, a magnifying glass, a calculator, a ball bearing for quick floor level check, a pocket knife and a hunting knife. When the situation warrants, some personal defensive protection should be included.

After assembling the required tools for an investigation, the first item on the itinerary is to obtain access to the site. Arrangements usually are made through the client for obtaining keys, etc., and/or permission to enter the area; often the client will want to accompany the expert during the inspection. There are occasions when the site will be unoccupied and, other times, when it will be crowded with workers. There are also times when the site management may be less than friendly – although minimally cooperative, depending upon which side of the issue the expert represents.

The proper way to inspect a site is to be thoroughly professional and adequately prepared to get the job done in the least amount of time. Preparation means having studied the relevant information and being familiar with the layout, construction, and any special conditions of the problem to be addressed. It is also helpful to take along a line drawing of the area under consideration clearly depicting plan layout and compass orientation. On some occasions, when other areas of related expertise are desired, it is advisable to bring along qualified experts to conduct their investigations at the same time; their reports will become appendices to the expert's final report providing back-up information as necessary.

As an example, when I am studying a building foundation-settlement problem, I often bring a geotechnical engineer for assistance to supply soil-related information to aid in my study. The direction the search should take is to study the scene of the problem carefully, duly recording the problem area, the time occurrence, the date, weather, temperature, condition of the area before and immediately after the problem, and the nature and extent of damage suffered. This information will provide the background for constructing an outline of the sequence of events. All information collected should be documented, even though it may appear irrelevant at the time.

The second important function in visiting a site is to record everything that may be pertinent to the problem. Memory will surely fail, and subsequent trips, as mentioned above, may not be available because of legal considerations. Information missed, forgotten, or unclear may prevent arriving at the proper conclusion - or worse, could be the deciding factor in the outcome of a case. Often, experts are called during an emergency. While it may be imperative to reach the problem site as soon as possible, the expert must take a few minutes to learn from the caller the scope of the problem and then to gather his wits as to the items that he must take along to accomplish his investigation. In emergencies, conditions can change fast through the efforts of clean-up crews brought in shortly after an occurrence. Their efforts can obscure or even remove valuable facts as to the cause of the problem. Once at the site, however, all items bearing on the issue must be gathered patiently and painstakingly; the quest is to seek the truth of the matter, and the investigation must not be unduly rushed in the process. The camera is extremely important for recording the scene in emergency conditions.

The guide for the investigation is the itinerary made at the office and is immediately put into effect. If the expert will stick to this planned approach, very little will be missed. It also gives him the opportunity to add new information as things become more apparent. It is well to mention at this point that the expert should be dressed appropriately for the occasion. If at a construction site, dress in working clothes, in an office building, dress in casual attire. Experts are judged by others on how they appear and are treated accordingly. Thus, wearing a business suit to a construction site may put others on the defensive and less responsive to questions than if casual clothes similar to their own are worn. The same reception is usually met when visiting an office building in work clothes. Dress accordingly.

Once arriving at a site, again using a construction-related problem example, the following sequence of events has proven to be the most effective means of gathering information. The expert must keep in mind to be thorough, as access may be denied for a return visit at a later date.

1. The first act, after arriving at the scene of the problem and cordially greeting the hosts, is to take photographs giving an overall perspective of the building so it can be readily identified – including compass directions. This is followed by localized photographs of the immediate problem area concluding with detailed photos of the items of concern. Because photographs instantly record a scene as it exists at the time of the inspection, there is no other substitute for this kind of record. Video cameras, as mentioned above, are also an acceptable method of photographing but are somewhat cumbersome to present in a courtroom unless edited to show only pertinent data. Photograph all angles to get a clear graphic representation of the conditions at that time. If photographs were taken by others prior to the visit, he should try to get copies for the file to add to his

47

collection; in so doing, be sure to identify the photographer. Photographs should be accurately identified as they relate to the specific problem addressed. Extra film is a must.

All data should be gathered that pertains to the problem. The expert should measure, verify, and record conditions of construction materials, details, sizes, dimensions, types, grades, etc. He should use the tape recorder to record descriptions and thoughts about the condition of items as he goes through the itinerary; also, he should make necessary sketches to items seen to recall his thinking at the time of inspection.

2. When interviewing witnesses, he should interview all available parties to the problem, witnesses familiar with the site conditions, eye witnesses, and the like. He should obtain permission to talk to witnesses from the employers - the representative of the top management of the company, not supervisors. Supervisors may balk thinking their jobs may be in jeopardy because of it. His demeanor during questioning must be professional, never humble or patronizing. Also, he should not take for granted any witnesses' statements or references without subsequent verification. He should use the tape recorder during interviews, but only with permission. Questioning should probe into the history of the site, on-going problems, conditions contributory to the problem, the time of day, weather, temperature, and any information that might assist in shedding light on the cause. He should probe thoroughly into the situation as it existed at the time of the problem.

    Some major points on interviewing witnesses are to always be professional, friendly, empathetic but not sympathetic, asking only those questions that are pertinent to the quest for information. He should give no opinions or answer any questions about his thoughts or conclusions. If asked, he should be noncommittal saying he hasn't formed an opinion as yet, that he's still in the process of gathering information. Regardless of the informality of the situation, he must keep his thoughts under wraps and concentrate on the task of obtaining information. I find it productive to pre-think a list of possible questions to ask witnesses. This prevents having to think them up extemporaneously, although some will come during the conversation.

On some occasions, he will have the opportunity, as in the case of working with lawyers, to review depositions of parties involved in the problem prior to visiting the site. Having done so, it will be of significant advantage in interviewing these witnesses during his visit. When interviewing, he should be sure to obtain names, addresses and telephone numbers – if possible, their positions, locations of work, and any other observations as to character, emotional condition, involvement in the problem, etc. Because the expert arrives after the fact, depending upon the situation, the client himself may have been present when the problem occurred. The expert should be cautious in discussions with clients as they are sometimes quite biased and will press for a favorable opinion, even if it means stretching the truth somewhat.

Finally, the expert should always keep his emotions under control regardless of personal feelings. I once questioned a foreman about an accident that caused the death of a workman, the evidence gathered pointed to his negligence as the cause. It was difficult to be civil to the man. Case in point: always be objective, always be neutral, always be emotionally calm. It is the lawyer's job to call for justice.

3. Check the weather bureau for climatic conditions on the date and time of the problem, if applicable.

4. Prior to departing, he should make a thorough check of his itinerary of items along with any others added during the inspection. Label and date all sketches, photograph film rolls, Polaroids, etc. Be satisfied that all items have been answered as much as possible. When I have completed my task, I find it beneficial to pause and reflect on the overall situation. After hours of searching and recording detailed technical information, the mind has a tendency to wander. By taking a brief rest and just looking around, often some overlooked information may be found. Don't be in a hurry to leave, even if pressed.

5. A final word about the site inspection. The expert should be aware, when inspecting a problem with a possible safety hazard, it is quite easy to become so wrapped up in the investigation that his own safety may be put in jeopardy. I was once inspecting a partially collapsed building where I became so engrossed in my study, I unwittingly placed myself in a very dangerous position.

Once all data has been gathered, a return to the office for transcribing recorded notes, tagging, and dating information is in order. It is best if this can be done immediately while everything is fresh in the mind. This new information adds more pieces to the problem puzzle. Once in place, they may show the necessity for further research and study, additional calculations, or more witness queries in order to put the final touches on the information to begin the analysis. In addition to the investigation data, he should keep a time and date log of all meetings and telephone calls in chronological order pertaining to the case. If deposed or testifying in court, the expert will be asked to recall the time and dates of all visits. It is exceedingly difficult to remember exact dates and times of past events, do not depend upon memory. It is also a good record for billing purposes.

The first rule of analysis is never to assume anything. Just as in the site investigation, the expert should have a systematic plan for reviewing all gathered information, performing necessary study, research, code requirements, calculations, and applying skill, knowledge, and experience to arrive at the most probable cause(s) of the problem. In cases of workmanship or defective materials, the guiding phrases are: (1) "Does the condition, as constructed, meet industry standards and, (2) Does it meet the accepted standard of care?" These two criteria are the most important items to use as judgment factors in determining liability in cases involving defective construction. After reaching his conclusions, he also may be required to produce several acceptable solutions for repair of the damage, accompanied by a cost estimate.

Often, it is impossible to collect all relevant information at the same time. During the investigation process, which may be as short as an hour or as long as several weeks or months, new information can trickle in from the client, or others, which may alter previous conclusions. Opinions should not be finalized until all possible information has been gathered - and the expert will never get it all, there is usually something in legal proceedings that appears from the opposing side just before trial begins. The expert should be flexible in his thinking and willing to take into consideration any new information that may have bearing on his previous opinions. His purpose has been to find the truth of a problem and present his thoughts based on the information gathered during the investigative process.

Robert J. Crawford

The collected information should now give a reasonably clear picture of the cause of the problem as well as what needs to be done in the way of correction. Sometimes things fall into place, at others, it requires a great deal of hypothesizing to arrive at the most probable way the problem could have occurred. In such cases, the answer is determined by the process of elimination, the most plausible being the chosen result. In any event, the expert now arrives at his opinions and conclusions based upon the information obtained and by applying his knowledge and experience in the field. The case file containing the transcribed notes made during the inspection and the summation of his analysis now forms the basis for the written report. This report, when completed and signed, becomes his legal testimony and may be used in court. In closing this chapter, the expert should treat every investigation, regardless of size or simplicity, as if it were going to court; that may be just what will happen.

CHAPTER 4: THE REPORT

Since the beginning of the written word, the advice of the ages has been to record thoughts and agreements in writing if they are to be remembered accurately and without dispute. Nowhere in the arena of legal activity is it more important to set down in writing your opinions and conclusions than in investigative work. Memories fade with time; what is positive today will be doubtful tomorrow.

The purpose of this Chapter is to guide the expert through a logical method for putting thoughts in effective written form. Whether you are outlining the information for your own use or compiling the report for others, a clear record of the facts when they are fresh in your mind will be of great assistance should the necessity arise to protect your position. The value of this can readily be seen in watching a court proceeding when, under questioning, the witness struggles to recall some incident of several years before. Every important point undertaken in the investigation must be reduced to clear, understandable language. In creating the report, the writer must take into consideration not only those for whom it was intended but also anyone who may have an interest in what is contained therein.

A common mistake made by many experts is to present their findings in the vernacular of their craft. While those conversant with the field readily understand, lay people do not. Therefore, it is important to choose words carefully so even those with limited understanding will grasp what you are saying with a reading of the report. It is well to remember, juries are composed of the citizenry; descriptive technical words will only serve to confuse them, thereby placing an added burden upon the attorney and the expert for further explanation.

While the report serves as valuable information to the client for decisions on his part, the usual case is that it provides the basis for legal action by attorneys acting on their client's behalf. It is also important to realize that the choice of words one uses in composing the report can have a significant impact upon those who read it. In preparing for trial, the attorney looks to the information contained in the report for points that will enhance his case. Conversely, the opposing attorney - as well as his own experts - will review the report for any items that may lead to a discrediting of the author. Words, therefore, must be carefully chosen to convey their exact meaning. The information presented must be so worded that the facts and resulting conclusions will be understood in their proper context.

Sharp attorneys can take as long as two hours to query an expert about a one page report. They literally dissect each word by questioning its meaning and relevancy in context. A common mistake in report writing is to use words that paint a situation either black or white when the answer really lies somewhere in between. Such words as "always," "never," "definitely," "without doubt," etc. are normally used lightly, but they are actually saying - and will be interpreted as - the condition addressed could never be otherwise. Most opposing attorneys will jump at such definite statements looking for holes in the expert's testimony. Use words such as "appears," "possibly," "seemingly," etc., as they express a thought about a condition but are in reality noncommittal. If you are positive about a point, say so, but don't be caught up in a definitive statement unless you are certain of your ground.

The report begins as a collection of thoughts, previously written notes, etc. gathered during the investigation and dwelled upon during the analytical period. It is a great help if the writer has the ability to express his words through the skill of typing. The computer/word processor is an excellent tool to use in setting down the information. It offers the writer the luxury of editing his statements during the typing process and, as related thoughts on the subject come to mind, they can easily be incorporated. If such typing skills are lacking, a good alternative is to dictate into a tape recorder. The writer can then give a narration of the events in the manner necessary for constructing the report. The recordings are then typed into rough drafts for study and editing.

In composing the report, keep items of information sequenced into a logical and orderly presentation. The length of the report varies with the subject and its complexity. Some experts compile many pages of text, lists, photographs, diagrams, drawings, and so forth. In some cases, this is a necessity. In others, unfortunately, most of the report is "boiler plate" to create an impressive package for the justification of fees. Reports of this type are usually difficult to read in trying to extract specific information. Reports also are compiled in the form of a few brief descriptive statements followed by long itemized lists of conditions discovered. Such reports make for difficult understanding as they present a group of facts and conditions without explanation and nothing to tie the data together.

A recent written report depicting the examination of a building by a registered engineer illustrates this point. His findings in the various parts of the building were presented in paragraphs as a jumble of nonrelated sentences. There was no continuity of thought as he jumped from one topic to another. It was very difficult to keep track of his thinking; I ended up dissecting each paragraph, collected each of his statements, and then placed them into appropriate groups. I doubt that anyone reading his report was able to understand it without going through the same process.

Reports take various forms, depending upon the author's writing ability and method of presentation. Many experts use a category approach dividing the information into sections depicting the various phases of the case. Each portion states the project involved, the background information on the parties to the action, the description of the building and its history, the problems to address, and, finally, their analysis and conclusions to the investigation. Others write a narrative approach along the same lines creating a commentary similar to a lecture on the subject. Then there are those who fall somewhere in between. It makes little difference as to the form it takes as long as it contains the required information and is easily understood by all who read it.

The most effective report is brief, concise, well written and to the point, using only enough information to state the facts in the case, illustrate the points, the opinions and conclusions and, when necessary, recommendations. Graphic data should be used if it serves to enhance a point or reference given in the text and should be cross-referenced. Eliminate all unnecessary words and unsubstantiated opinions, they serve little purpose and supply the opposition with additional information to attempt to discredit your findings. On the other hand, however, do be thorough in presenting detailed information necessary to substantiate your position. Thoroughness is required, but brevity is the goal.

Summarizing, the expert must be able to effectively communicate his thoughts. He must present an accurate record of his findings, analyses, opinions, and conclusions in clear, precise English using words easily understood by the lay public. He must also possess the ability to orally elaborate on his written work when required to present the information to others. Such would be the case during testimony in mediations, depositions, or at trial.

One very important factor in report writing is the maintenance of the expert's file. In Chapter 3, we discussed the origination of the file and the data contained therein. The report is based upon this information and, once the file is completed, it legally becomes discoverable evidence when the author is designated as an expert witness for the case. At that time, both the report and the file are attainable by parties of the opposition. When your client is an attorney, it is of the utmost importance that the report be written only upon his authorization and direction as there may be evidence within the file that would be detrimental to his case and his client's interests. Remember, the task of the expert is to investigate the problem and present his findings upon request; what is done with the information is the attorney's job.

With these thoughts in mind, we will now look at an effective manner in which to present the information contained in the file.

ESSENCE OF THE REPORT. A procedure is required to set down the thoughts of the expert in a logical arrangement that can be easily understood and fully informative of the entire undertaking. The following outline expresses the contents of the report in what I have found to be the most efficient and comprehensive format:

1. Introduction
2. Documents
3. Site description
4. Site conditions and investigation
5. Witness interviews
6. Analysis, opinions, conclusions, and recommendations
7. Summary
8. Closing statement
9. Appendices
10. Disclaimer

1. INTRODUCTION. Why is the report to be written? This is the first question that should be asked and its answer will give the purpose for the report. Be it a summary of events, thoughts as of a certain date, or information in response to a request from a client, the purpose should be clearly stated. If it is to document an occurrence that is self protective, as if the author himself has the problem, he should create a narrative that describes such in detail and includes the reasons why documentation is recorded at this particular time and date.

   In response to a client's request, it is best to start at the beginning. The report should be dated, sent in the name of the client to his address, referencing the building or site and its location. It should be in standard letter form written on good quality letterhead stationery. The opening statement should be directed to the referenced subject and state that the report is made in response to the client's request. It then states the client's relationship to the situation: owner, attorney, etc., the information desired, and the date of the request. If practical, add any additional comments that will elaborate on the reason why he desires the information and your instructions in the matter.

2. DOCUMENTS. Once the reasons have been presented, move on to indexing all written information that has been received regarding the situation and from whom. Document all items, the nature of the contents, including names and dates pertinent to the situation. They may take the form of plans, calculations, photographs, reports by others, depositions, and the like. In the case of plans, record sheet numbers and be sure that they have appropriate authenticity and signatures, such as engineer's stamps, signatures, building permit stamps,

dates, etc., for validation. Documents without official stamps, etc. are easily discredited for lack thereof. Be brief and accurate in your descriptions.

It is extremely important to record all data and to keep the copies given to you for study, if possible. They will contain your markings, notes, etc., which will indicate that you have examined them in the course of your investigation as well as the source of the conclusions you have drawn. If you are unable to keep the materials, initial each piece and then have photocopies made indicating proof of authenticity. It is important to remember that you will be basing most of your findings, analysis, opinions, and conclusions on the data furnished to you. If you are unable to positively identify the materials with which you worked in the case, it could be dismissed during a court proceeding thereby eliminating your testimony as an expert on that subject.

Another situation that can occur is where the information received was not correct. In a case such as this, the expert's testimony, with regard to the documentation, would not be usable in court though it would not discredit him as an expert. Any other opinions gained from the investigation would still be valid as long as it wasn't dependent on the faulty documentation. Make every effort to be assured the information received is correct.

As an example, on one occasion I received three sets of plans to study for a problem with a building. Each set had different dates, the second and third sets being subsequent revisions of the first. My analysis, opinions, and conclusions were based on these three sets of drawings. At the trial, a fourth set of revisions appeared and was claimed to be the one under which the building was constructed. My findings, only based upon the three sets, were then judged irrelevant due to not having the forth set. This did not, however, detract from my standing as an expert, it only pointed out that those with whom I worked did not supply me with the correct set of drawings. Subsequently, we found that the opposition had the fourth set but did not disclose it until the trial. In that particular case, however, the builder did not construct the building properly from the fourth set; my site inspection uncovered code violations in the completed building regardless of the information contained in the fourth set of plans. It is extremely important to be assured that you have correct information on which to base your studies. Your client should be queried as to the authenticity of the information.

3. SITE DESCRIPTION. Give the time and date for the first and, if applicable, subsequent visits. Include the location, address, names and positions of those to whom you spoke and others who may have accompanied you on the tour. Also, include whether access was granted cordially and if you were allowed to view all areas of the site pertinent to your study. Record the weather, temperature, wind conditions, etc., as necessary at the time of inspection. If the problem occurs at a building, describe the site, location on the site, size, number of stories, general layout, construction type, material composition, and any special features or conditions noted during your visit. While the overall perspective acquaints the reader with the general scene, be specific in describing the conditions with which you are concerned.

4. SITE CONDITIONS. Once the general layout has been covered, it is now time for detail. This is a zeroing in on the problem that you are to address. State those items that are pertinent to your study and describe what you saw in regard to the problem. List each component of the conditions found, any subsequent problems caused by the conditions, what has been affected, and the current state of affairs. State measurements taken, discrepancies found, and comparisons with plans and other drawings. If the scene had been changed since the problem occurred before your investigation, describe in what manner. Make sketches as necessary.

5. WITNESS INTERVIEWS. On those occasions where you have the opportunity to talk to witnesses, their recorded statements should be studied for information that can be used in the report to shed light on the conditions as they existed at the time the problem took place. The credibility of a witness' remarks is a matter of judgement, but should be included if it is pertinent to and will give a greater understanding in your analysis of the matter. Be sure to obtain names, addresses, and telephone numbers of all witnesses interviewed.

6. ANALYSIS, OPINIONS, CONCLUSIONS, AND RECOMMENDATIONS. The analysis phase of the report is a narrative description of how you viewed the problem, the apparent causes that led up to it, parties that were involved, the effects caused by the problem, and possible solutions. It states the path you took to uncover the evidence and how you applied your knowledge and experience in the field to determine the cause and effect of the problem. These are based upon your observations, knowledge of materials, construction methods and practices, analytical calculations, and knowledge of applicable codes and ordinances.

61

The analysis is complete when you have formed your opinions and conclusions in the matter. These should be presented in detail, stating your reasons for such and upon which of the many pieces of information gathered were the deciding points. Such points will be dwelled upon by both your client and any opposition that may come into play in the future.

Often experts are asked for recommendations for repairs, etc. to solve the problem. I have found that there are usually several ways in which repairs can be accomplished. The question is which is best, taking into account the essence of the problem, timeliness, and the costs of repair. If, as a matter of course, you also provide such services, it would be helpful to make a statement of your availability and willingness to do so in the report, if so inclined. I usually advise my clients as to the avenues of repair, stating the advantages and disadvantages of each and let them make up their own minds. If you are unable to offer such assistance, it is usually good personal relations to make recommendations of qualified contractors or craftsmen to do the work.

7. SUMMARY. Reports are often difficult to fully comprehend because of the significant amounts of detailed, and often technical, information that is required. It is difficult to keep all the relevant facts in mind, as one reads on, gathering yet more information before coming to the conclusions.

I have found that clients appreciate the report to end with a short paragraph that summarizes very clearly and simply what has been said. It capsulizes the effort in a few easily understood statements giving the essence of the problem, pertinent facts, the cause, the effect, the main reasons deduced from analysis, and finally the opinions and recommendations. It concludes, if applicable, with a statement as to the best of several alternative solutions for repairs. Such a summary saves time and effort and says it all in one place. If further elaboration on details were desired, then a reading of the entire report would be in order.

8. CLOSING. The report should end with a qualifying statement that it was written in response to and within the guidelines and extent of the client's request. Often a client will want only a cursory examination, one that can be performed with a brief walk-through inspection. When such is the case, you should make a statement of the type of inspection made and disclaim any further exploration into the problem. If you find that further investigation is needed because of some apparent problem discovered during the process, you should so note in your report and recommend that such be done to determine the full extent of the problem.

The report should always be signed with a wet signature and your professional stamp with license number and expiration date. Copies of the report should be given only to those individuals whom the client has designated for receipt, such information being deemed confidential.

9. APPENDICES. To assist the credibility of your report, you should make available to the recipient all items of information listed in the Document section that were used as a basis of your efforts. This information is usually the data received at the beginning of the investigation and any pertinent written information collected in the interim. Additional information could be any subsequent written correspondence, reports,

subsequent depositions, calculations, diagrams, drawings, photographs and the like that will substantiate your findings.

10. DISCLAIMER. This is a short paragraph limiting the involvement of the expert in the investigation. It is especially important to state that any problems in the case, other than those expressly addressed in the Report, are not a part of the expert's work product and responsibility for such is excluded from the study.

The expert's written report is a legal document that, when signed, can be used in a court of law. It is the professional opinion of its writer and expresses his knowledge, expertise, and experience in reviewing the problem. Great care should be exercised in its writing to express only those opinions that have been formed on the basis of his study. Statements that are open-ended or ambiguous should be avoided. Once the report is completed, several copies should be retained in the file. The file should be edited, and all nonessential information to the case should be discarded. All remaining items need to be organized such that, with a brief review, one may be brought quickly up to date and gain a thorough understanding of its contents. This is especially important because if the information is to be used in a trial situation, several years will probably elapse before such comes to pass.

The standard letter form is used for the report as is customary in business proceedings. If it contains numerous pages, it should be in book presentation form. The report is representative of the expert in absence and will be judged by others upon its presentation and information contained therein.

CHAPTER 5: THE ATTORNEYS

Our attention now turns to the legal process, which is an integral part of the resolution of problems. When disagreements occur between parties that cannot be resolved amicably, the remaining avenue of resolution lies in the legal processes – arbitration, mediation, or the courts.

Disagreements between parties and their resolutions are the realm of the attorney. Such occurs in all phases of business and personal life; here, we are concerned with issues involving the use of experts. Because attorneys are at the heart of the process, some understanding of their functions and expertise, as well as their wide varieties of personalities, is in order.

Attorneys are generally intelligent, well educated, and quite proficient at their trade. There are exceptions, of course, but in the main, they handle their tasks with competence and integrity. While there are many jokes and disparaging remarks made at their expense, they are an essential part of the realm we term as society. They are prevalent in politics, all parts of government, and throughout our free enterprise system - perhaps too much so, in the opinion of some. The legal profession is presently self-governing, that is, they police their own members with respect to licensing, etc., through their American Bar Association. The other professions, such as medicine, accounting, engineering, etc., are regulated by the States of residence. Attorneys are well paid for their services, often receiving fees upwards of 40% of court-awarded judgements - which is a concern of many who end up on both sides of an outcome. Our task here, however, is not to dispute the way things exist but to understand how they work and the expert's role in the picture.

Attorneys perform numerous duties in our society. The practice of law ranges from constitutional matters to litigation.

For our purposes, we are concerned with the practice of litigation, as performed by trial attorneys. While the courts pass judgement on both civil and criminal cases, our interest lies in civil suits. Judgement, however, is rendered in several jurisdictions besides the courtroom; there are also mediations and arbitration of various types. The processes of each will be discussed in a later chapter.

When a dispute arises, there are usually two major parties to the conflict - the one who claims he's been wronged and the other that caused him the grief. The first step normally taken when such disagreements occur is a meeting of the opposing parties where their grievances could be aired and, hopefully, settled in an amicable manner. Failing this, the complainant party proceeds on to the courts for restitution. This brings the attorneys to the scene. Each party selects an attorney to represent him in the action. The initiating party, called the Plaintiff, engages his attorney who is known as Counsel for the Plaintiff. The opposing party then becomes the Defendant, and his attorney is known as Counsel for the Defense.

Upon hearing the problem and being assured it has merit and actual damage has been done, the attorney for the plaintiff prepares documents for filing of a lawsuit on the plaintiff's behalf. The lawsuit cites the various claims of wrongdoing inflicted by the defendant; it will also include monetary damages required to correct the claims, plus attorney's fees. The plaintiff's attorney's first task is to gather as much information about his client's problem as is available in order to document the list of claims involved in the suit. A significant part of his task is to engage competent experts to investigate the claims and document their findings regarding issues. Once the attorney has gathered enough information for substantiation, he creates his plan of action for pursuing the case. A large part of the plan is consultation with his experts. He needs them for understanding the significance of items in the problem area and to get a grasp on the intricacies of their studies that lead to their conclusions and opinions. During these meetings, he also coaches his experts on how best to present their information in preparation for depositions and trial. Once the plan has been completed – and updated with last minute changes - he is ready for meeting the opposition for settlement purposes through mediation or arbitration; if unsuccessful, then on to trial.

After the plaintiff's attorney has completed his plan, the defendant will then be served with a subpoena informing him that he is being sued. Sometimes this news comes as a shock, especially if there has been no previous indication of such intent. However, in our scenario, the defendant was aware that it would be forthcoming. Upon receipt, the defendant engages an attorney to represent him in the action. As the defendant's Counsel, he comes onto the scene significantly later than does Counsel for the plaintiff. At this point, he follows much the same procedure as his opponent in collecting information and lining up experts for assisting in the quest for information. He is aware the investigation for substantiation of the claims has been on-going by the plaintiff's experts for some time. He now desires this information to know the basis of their claims, as described in the lawsuit. He then proceeds to obtain that information during the deposition process. So informed, he then engages his own experts to investigate the problem and report back with their conclusions and opinions. Thus, knowing the claims against his client and the findings of his own experts, he can formulate his plan for the defense of the lawsuit.

Once the defending Counsel's experts have been engaged and completed their investigations and reports, the plaintiff's attorney has the opportunity to deposes them to gain their opinions and recommendations for their side of the case.

Thus each side has a team leader in their attorney; it is his task to develop a game plan to prosecute or defend his case. For the expert, it is very important to understand how attorneys think, how they perform their roles, and the tools they employ to extract information from those whom they question. As noted above, most are very adroit in their highly developed technique of questioning - and thinking on their feet - to obtain the maximum amount of information to aid their case. They are also well schooled in the use of emotion and timing during depositions and cross-examinations to achieve their goals, as will be discussed later.

In regard to law firms, look into the Yellow Pages of your telephone book under Attorneys and you will be surprised at the number of legal practitioners in your area. It is difficult to believe that so many are actively engaged in the community's business atmosphere, especially at the fees they command. Yet the statistics indicate the profession of law has more practicing members, percentage-wise, than any other professional field of endeavor. The demand for legal action by the populace has risen dramatically in the last several decades - which gives an indication of the answer. Regardless of the number, most seem to be gainfully employed and busy at their trade.

As in most businesses, there are those of long-standing and prestige in the community; law firms are no exception. Some of the more prominent firms have forty to several hundred lawyers, depending upon the size of the city, and occupy the top floors of the largest office buildings in the business sector. Their legal departments cover all phases of law practice. Other firms are in the middle range with ten to forty attorneys, then, on down to those composed of several partners and, finally, the sole practitioner.

Because experts assist the legal profession with their tasks, most contacts are with the litigation attorneys. These are the people who require information outside of their own expertise. They know the law, but most are at a loss when it comes to the technical aspects of other trades or professions. There are some members of the legal profession who try to give the impression that they are well informed in an expert's field during cross-examination query, but most have only a superficial knowledge of a given subject - on which they crammed just prior to the questioning. However, some are indeed well versed in particular fields, such as construction. I had the experience of working with a litigation attorney who was also a registered Civil Engineer, we talked the same language, communicated effectively on the issues, and the value was evident at trial.

As experts who work with attorneys, it is particularly useful to gain some insight into their personalities. It is especially helpful to be able to recognize and distinguish a given type when he is the opposing attorney. Knowing the signs and tactics to expect goes a long way to providing a good defense against such adversaries.

First of all, some generalities. Every practitioner of the legal profession has a very good impression of himself. He is well educated, wears expensive clothes, usually works in one of the more elegant office buildings, drives a prestige car, earns a fine salary, and is given high status by others. Their personalities vary from quiet and polite to loud and boisterous - no different than other professional people. I personally enjoy working with attorneys as they are interesting, usually friendly and courteous, quick to comprehend, and are generally a challenge to educate in my field in preparation for trial. Below, I have classified several of the more interesting types I've met in my experience during cross-examination questioning and will introduce them to you along with my comments:

1. THE BULLDOZER. This attorney will come at the expert like a runaway earthmover. His approach is to intimidate the expert into supplying the desired information by appearing to be infinitely strong and the

ultimate in authority. He will often question – in demanding and challenging tones – every single facet of testimony even though the answers are readily plain to everyone in attendance. This type tries to intimidate the expert into submission, thus causing doubt about his own testimony. The best defense to this strategy is to just ignore the assault, remain calm, and give answers to the questions in a normal, confident voice. Treat him as an undisciplined child with great tolerance, which lets him know that you aren't buying what he's selling. Once he sees that he cannot intimidate you, he will usually change his tactics to something else. If this attitude occurs during a deposition, ask your attorney to object to his manner of questioning during a break period.

2. THE FRIEND. This attorney, through the whole questioning process, will give the appearance of being a "friend" who just wants a little information about your findings and opinions. Most of his questions are mild, and he gives the impression that he fully appreciates your cooperation in this matter. Then, when he says he has completed the deposition, he rises to leave. At this time, when the expert feels that the session is over and begins to relax, the attorney suddenly remembers that he has just one more question and comes back with a knockout punch that takes the expert totally by surprise. Such a tactic can cause him to drop his guard and, shocked by the sudden challenge, reveal information without thinking it through, which may be damaging or inconsistent with his prior testimonial opinions. The questioning will then continue in that vein until the attorney is satisfied he's pried out the information he wants. The best defense against this type of attack is to keep in mind that you are in the enemy's camp from the minute you enter until the door closes behind you.

70

Never let your guard down while in the presence of an opposing attorney.

3. THE MARATHONER. This attorney will question the expert endlessly on every possible aspect of his report, previous testimony, background, education, conclusions, opinions, recommendations, ad infinitum. His tactic is to wear the expert down to the point where he is weary from all the questioning, drops his guard, and reveals information because of exhaustion and just wanting to end the proceedings. These are the marathon attorneys and they are usually well-funded to go to the added expense. My experience with one of these fellows was in a deposition. The case was uncomplicated, regarding a roof failure problem and its subsequent repair. I was assisting the plaintiff's attorney; the opposing attorney was representing an insurance company. The deposition began on a Monday morning at 9:00 a.m. and concluded on Thursday at 4:00 p.m. Four full days of deposition; I was asked more irrelevant and repeat questions than ever experienced previously. His client was an insurance company, of course, and his meter was also running for the four days. The best defense against this type is to be sure to get your eight hours sleep each night, request frequent breaks, and know your material.

4. THE CHALLENGER. This type of attorney is more common and you usually find one in every group of attorneys at a deposition. I happen to meet this one at a binding arbitration. His game plan was to literally attack every statement I made regarding my findings during my investigation, as contained in my report. He took the stance of requiring my complete definition or explanation of each item questioned. A typical example was my statement that a particular defect in the building was a violation of the Uniform Building Code. He

71

demanded I give him a definition of the Code, its relevance and purpose, what it contained, how it was relevant to the particular issue under consideration, and so on. He did the same when I described another defect in regard to an undersized nail, which was typical of those used in the building. All during his questioning he was mostly hostile as he tried to intimidate me with his aside remarks. The best way to confront this type of adversary is to treat him civilly by being cordial, answering each question asked, no matter how insulting, in a calm manner completely ignoring his ploy. The Arbitrator, who was visibly irritated by his attitude, didn't buy into his defense either; he lost.

5. THE PERFORMER. This type of attorney is quite entertaining. To add dimension to his oral presentation, he displays physical actions that are intended to add impact and drama to his questioning. I met this one while testifying in court. He was representing a fellow who climbed outside a second story balcony guard rail to make a repair and, while leaning outward and hanging on with one hand, fell and broke his leg. The attorney had a mock-up partial model of the balcony and guardrail in court for demonstration. With a voice that was supposed to simulate the best of a "Lawrence Oliver," he hopped upon the raised platform edge with one foot, grabbed the handrail with one hand, and started swinging back and forth continuing his lament of his client's injuries all because of the supposed faulty construction. It was quite a show. The jury was entertained but didn't buy his story. I was amused, too but didn't divert from my rehearsed composure and attention to the case at hand. Always be prepared for anything during deposition or at trial.

6. THE GENTLEMAN. Most of the attorneys I have worked with or been deposed by have been polite, cordial, concentrated on the business at hand, and were not given to the tactics of those above. The Gentleman knows his business and goes about it with dispatch. He stays to his game plan and, within the bounds of professional conduct, does his best to pry information from the expert to aid his case. He doesn't use gimmicks, he uses his brains and talent to follow leads that are uncovered in the proceedings. Because he is very adept at his craft, this is one attorney with whom the expert must be especially on guard. Again, the best defense is to know your material, keep your wits about you, remain poised, calm, and answer the questions to the best of your knowledge in short, precise replies.

Working as an expert witness within the legal system is a process of assisting attorneys in their practices, and it takes two very different directions. The first is aiding the attorney in preparing his case, the other is presenting oneself for examination by the opposing attorneys.

To be of assistance to an attorney, the first, and most important item you must determine, is if his problem lies in his area of expertise. Some attorneys will take any type of case, even though they have little or no experience in that field. It is a difficult task for an expert to have to completely educate the attorney in his field, and it is especially dangerous during the infighting in the legal arena if he hasn't the depth to really understand the expert's information. This is the reason why it is good practice to query the attorney on his knowledge in the field of the problem.

Calls for assistance from attorneys come through efforts of advertising, prior acquaintances, or the most usual course, from referrals. During the initial conversation, the first order of business is to determine exactly who is the caller, the law firm he is representing - along with addresses and telephone numbers, and then an accurate description of the problem in which he is engaged. This preliminary information will give a quick summary to form the basis of deciding if you are able to assist him in the case. More importantly, it will also allow you to judge whether you want to offer your assistance taking into account other criteria, such as conflicts of interest, personal relationships, or friendships with the opponents or their experts. I make it a practice in my work to avoid opposing other experts who are personal friends or professional or business colleagues of close proximity. I do assist in cases where a friend or professional acquaintance is in need of my services. However, impartiality must be maintained and the facts represented as they appear.

My initial queries concern the law firm and whether I have been in opposition to them in the past; sometimes your performance as an opposing witness in a case may impress them to desire your services. This is a judgement call. I recently turned down an offer from an attorney to assist on a large case because my prior experience with him as an adversary was such that I preferred not to become involved. I disapproved of his general approach to the investigation and his personal demeanor to those involved in the case. He was one of the Bulldozer types and desired my services for his creative "witch hunt." A little subsequent follow-up on the case proved my option was the right one as his second choice accepted the assignment and told me later that he wished he hadn't. So it pays to take a little time to get a reading on the prospective client, to learn the scope of the case and who may be involved.

Once you have determined that your participation may be beneficial to both the attorney and yourself, the next step is a personal interview and, if satisfactory, to begin the information gathering process as well as to finalize the fee schedule agreement. Personal impressions mean a lot; by meeting in person, you bring into play personal observations of the client, his office, and the state of his practice as reflected by the surroundings. One of the dangers to an expert's reputation is falling in league with unscrupulous clients. It is easy to become unwittingly involved with a less than honest client unless care is taken at the onset; subsequent resignation from a case may leave some unwanted ties to the situation.

Other opportunities to avoid are those cases that are lose/lose situations where, even if your side wins, it is still a loser because of the emotional or ethical issues involved in the case – especially if it receives prominent newspaper coverage. Such cases cast a pall on all people involved, and it certainly will not enhance your reputation in the field. The old saw about guarding your reputation like your daughter's morals is true. Other items to look into are such things as the attorney's experience: is he the new kid on the block or a seasoned professional? Is he a "make work" kind of lawyer who chases ambulances or takes any case that walks through his door? Is there true value of service in his case or is he assisting his client in chasing the dollar? If he appears to be a competent craftsman, is he easy to work with and does he have a professional manner of approach to his work? What is his modus operandi on his cases such as this? What experts has he used in the past? What were the problems with them, if any?

As an illustration, I was assisting a fairly seasoned attorney in a case where there were three other experts on our side in fields related to mine. During the course of the investigation, we met individually and collectively several times. I found them to be credible in their particular areas of expertise and was satisfied that they were assets to the cause. Each of these experts was deposed and eventually testified at trial. My turn to testify came last in the process. I had been working closely with the attorney, and he was quite free with his comments about the other experts.

After each completed his task, one by one, the attorney had said aloud how disappointed he was in their individual performances. After the first two had finished their parts, I began to suspect the problem was with the attorney and not his experts. When he berated the third expert's ability, I was wondering what he would say about my performance. I had gone over our presentation very carefully coaching him on questions to ask in direct examination and my forthcoming responses that substantiated my opinions in the case. To my surprise, he did not follow our plan but went off in an entirely different direction leaving me with my prepared information unsaid. I was surprised to say the least.

When I was excused by the Judge at the end of my testimony, such as it was, I felt the entire process was a waste of time. As is said, you win some and lose some, but this fellow went off course so completely that I concluded he was the problem, not his experts. Therefore, it is well worthwhile to do some inquiring about a potential attorney client. A little routine investigation into his case track record - through other attorney acquaintances - may shed some light on him as a client, his reputation, as well his record for timely payment of fees.

Assuming that you are satisfied with the client, the case has merit, that you can be of valued assistance in the matter, and the timing and fees are satisfactory, the next step is to try to gain as complete an understanding of the problem from the client as possible. As noted previously, attorneys vary in their knowledge of your field; therefore, it is imperative to determine the extent of his understanding of the problem. By the time attorneys engage an expert, they have already been told many things about the issue; some they understand, others they have only a general idea of the problem. The expert's job is to ferret out all information and put it in proper perspective for the attorney's understanding. Some defects, as in construction cases, may be obvious; however, their causes may not be and speculation can abound with uneducated guesses. It is the expert who will determine, according to his expertise, education, and experience, the actual basis in fact. Once determined, he will put his findings into appropriate context for his client's understanding.

Once the client has provided the expert with all available information - both oral and documentative, the expert's next task is to study the material to gain a grasp of the problem. He then creates his outline procedure for the investigation, as described in Chapter 3. On completion of the investigation and composing his report, as outlined in Chapter 4, the expert's task is to educate his client on the intricacies of his field and explain how the discovered information fits into the perspective of the case. He should meet with his client and explain that portion of his field in which the case is a part and how the collected information formed the basis of his conclusions and opinions. Once this understanding has been gained, the expert then assists in formulating questions to be directed at the opposing experts in accordance with the client's game plan. Often, if the client is a little unsure of his footing during the deposing of his opponent's experts, his expert may attend giving advice as to questions to be asked in response to answers received. Such advice may be given via written notes and/or confidential conversation. I have used this tactic a number of times to great advantage. It is especially useful if the opponent's expert is in over his head and is guessing, misstating published information, or the like.

In one recent case, this situation occurred: the deposed expert had not studied the code material affecting the line of questioning. He obviously did not know the answers and, unfortunately for him, began a guessing game. While trying to bluff his way through, I was quietly advising my client on the inappropriateness of his responses. This resulted in negating most of his testimony and credibility in the eyes of all attorneys present. A simple "I don't know" or "I don't remember" would have served him well except for that particular line of questioning. It is important to do your homework; if you miss something – and it's quite possible, considering the amount of information to be retained for questioning – just be honest and say you either don't know or that you don't remember but can look it up.

While the attorney-client is being coached and prepared by the expert, the expert himself is also being readied for his presentation. It is obviously a two-way street of information giving and sharing. The attorney and his expert are a team effort; their goal is to present their side of the case in as efficient and factual manner as possible in hopes of convincing those judging the proceedings that their client should prevail. It is a time-consuming and often a brain-wracking process, but it is an absolute necessity if the information gathered by the expert is to be expressed correctly at trial.

The shoe is on the other foot when the expert is being examined by opposing attorneys. While he may be an able assistant to his client in preparing his case, he must also continue this assistance while under fire by opposing counsel. As noted earlier, the discovery process is a useful tool in the attorney's kitbag. Simply put, discovery is the process by which an opposing attorney is allowed to query opponent experts to their heart's content about any issues they choose as related to the case, as well as any personal information concerning the expert's credentials into which they care to dig.

A major goal of the opposing attorney is to discredit or impeach adverse counsel's experts. This is done through a series of questions that explore the educational background and experience history of the expert; i.e., his education, work experience in his profession or craft, and his involvement in legal cases. He will dwell at great length on the issues of the current case utilizing questions based upon information supplied by his own experts, his knowledge of the issues, the documents prepared by adverse experts, and any additional information that crops up during questioning.

During the deposition process, the expert must be ready to do battle; however, it is a passive contest as waged by the expert. His task is to provide as little information, though accurate, to the opposition. He must present himself as a credible, knowledgeable, and honest practitioner in his field, as he should be. Because a period of several years usually passes before the issue comes to trial, preparation for this encounter is extremely important and includes the following:

a. Review all documented data; i.e., plans, calculations, exhibits, photographs, etc.

b. If possible, re-visit the site or place of occurrence for memory refresher; it is well to remember once again, you may not have a second opportunity to see the place and, therefore, your original photographs must be thorough enough for complete recall.

c. Thoroughly review your own deposition as this is the major tool of the opposition in trying to trip you up on previous statements. All answers must be consistent with prior testimony – unless conditions or new information has come to light and can be explained.

d. Review all depositions of the opposing side, paying particular attention to the overall scope, those involved, and their place in the scheme of things.

e. Thoroughly go over your file; be up to speed with the information as if it was documented only yesterday. It should be organized and indexed for quick reference. Remember, it will be copied by the opposing attorney during your deposition.

f. Prepare yourself physically and mentally, as outlined in Chapter 6.

I re-read Chapter 6 prior to every deposition I give; you cannot remember it all, so a refresher is a must if you are to perform at your optimum.

It is essential that experts, who are to undergo deposition and trial questioning, be aware of some of the tactics used by opposing attorneys. Litigation attorneys do not really know how good their experts are until they see them under fire. Because they obviously depend on their expert's knowledge, they rely on him to be well versed in the field of concern. Often, however, this is not the case. It is interesting and often amusing to be in attendance during a deposition and listen to some expert's testimony espouse theories that would make the members of their monthly Association meeting laugh. It is truly amazing to listen to the erroneous information put forth as fact when a cursory reading of the applicable sections of the pertinent code – or basic texts on the subject - clearly state otherwise. Attorneys also need to be careful in the selection of their experts.

The astute expert should always try to determine the type of attorney he is opposing to have an understanding of what to expect during cross-examination at trial. It is possible to get a good reading from the manner and content of questions asked during deposition. When he demands, for example, overly precise detailed information of a specific condition upon which an opinion was based - to the extent that it seems self-evident and even ridiculous to go further - his reason is usually because he has no idea what the expert is talking about. His tact is to keep him talking in hopes of obtaining some information with which he can get a hook into the expert's credibility. He is digging for gold hoping the expert will slip up so that he can catch him in subsequent crossfire at trial. The expert's best course is to just answer questions as plainly and directly as possible, keeping consistent with his findings and opinions. He shouldn't be rattled by this grilling; the harder that attorney probes, the less he knows and is in hopes of striking gold by ferreting out an off-guarded statement.

As you gain experience working with attorneys, there is one thing that will come into view. They have a mutual respect for each other even though they may be on opposite sides. Experts come and go, but attorneys in the same general locale eventually come head to head to do battle; at other times they are co-partners on the same case. In light of this, they usually have a common courtesy toward each other expressing due consideration while protecting their turf. This is the arena of lawyer work. Some have said they are akin to sharks having a mutual respect for each other - until one is wounded; at that time, the more lethal of the species will do his act. There are, of course, many highly ethical practitioners in the legal profession, but they are not seen as often as one would like. It is well to remember sharks may appear at the most unlikely times and often in the highest places.

In summary, disputes that cannot be settled amicably eventually move on to litigation between parties. The disagreement then enters the next phase as a legal battle between professional word warriors, known as attorneys. Each side in the dispute engages an attorney to represent his claims to be presented in court. The party that feels he's been wronged, the plaintiff, tells his story to his attorney – commonly called Counsel for the Plaintiff. The opposite party finds that he must defend himself against the claims and engages his own attorney, known as the Counsel for the Defense. Attorneys are people like everyone else with personalities to match. Most are intelligent, well mannered, liberally educated, and experienced in their field; personalities vary from the very intense to the easy-going types. However, all have a few qualities in common. They are very competent searchers for information, as this is the fodder with which they load their cannons for the attack or defense. They all have individual tactics for querying parties involved in the action, and their common goal is to discredit as much information from the opposing side as possible.

As each respective attorney gathers information to prepare his game plan, he often finds the need for expert knowledge in the matter at issue. He will then conduct a search for a person or persons who are knowledgeable in the specific field and hold proper credentials. The attorney usually relies on experts that he has engaged in the past or others recommended by close legal acquaintances. Most cases require the assistance of several experts working in harmony, preparing the information to be used by the attorney. The expert's job is usually twofold: (1) to assist the attorney in his education of the essentials of the particular field under consideration while acquainting him with the findings and opinions of the investigative process and (2), to prepare himself for the discovery process, as conducted by the opposing attorney. The expert's investigation that resulted in his discoveries and opinions must be clearly explained to his attorney. Following this, he must then coach his attorney on questions to ask the opposing experts during cross-examination. The expert should be well prepared to undergo a tedious examination of all aspects of his involvement in the case as well as to remember to avoid divulging information not specifically asked for by the attorney for the opposition. Working with attorneys is a two-phase job, experts have to be up to the task.

CHAPTER 6: THE DEPOSITION

The deposition is a legal tool by which opposing attorneys in a case are allowed to query principals and witnesses to gain pertinent information relating to the issue in dispute. This information gathering session is known as the discovery process. It is a powerful and useful tool in the legal briefcase where important facts, opinions, and conclusions of witnesses are brought forth for tactical consideration. During this discovery process, each side has the opportunity to size up the opposition, giving them thinking room for either settling the case or moving on to court. The discovered information is only one facet of the process; another is how effectively a witness performs his role thereby revealing his potential as an expert in trial.

Because deposition testimony is admissible in court, deposing attorneys will use it to gain a grasp of their opposition as well as to try to refute the expert's testimony during the trial by pointing out conflicting statements in hopes of bringing discredit or impeachment. Impeachment is a process whereby the opposing attorney can have a judge disqualify a witness because of lack of knowledge, bias, or conflicting statements. If, for some reason, the expert cannot be present in court, the signed deposition can be used as his sworn testimony. Thus, the deposition is a legal proceeding for which the expert should be fully prepared.

To give a meaningful deposition, the witness must: (1) establish himself as an expert in his field of expertise applicable to the case, (2) be confident in his findings, opinions, and conclusions, and (3) his dress and act in a professional manner. The latter is demonstrated by his personal appearance, the manner in which he presents himself, controls his reactions to questioning, his comprehension and directness in answering questions, and his command of verbal skills to communicate his thoughts.

The capabilities of deposing attorneys range from the very careless and unprepared to others who are highly skilled and knowledgeable in both law and witness examination. A competent attorney will quickly recognize the type of witness he has to confront. A major part of the legal game is to weaken, discredit, negate, or even disqualify an opposing expert. Therefore, it should be remembered that his main objective is to win for his client or, at the very least, minimize losses. Experts, therefore, are the opposing counsel's targets. During the pretrial period, both sides to an action are required to disclose their experts. This gives each an opportunity to size up the opposition. Sometimes an expert will act only as a consultant to an attorney in a case. As such, the information offered by the consultant is essentially advice and is confidential. However, once the consultant is designated as an expert, his advice, opinions, and conclusions become discoverable evidence by the opposing side.

The law provides a method for requiring the presence of an expert to be deposed; it is called a subpoena. This is a legal document personally delivered to the witness by a person employed to complete the task. Such messengers are commonly called Process Servers. The subpoena contains information denoting the witness to be deposed, the parties to the action, the deposing attorney, and instructions to appear at a given date, time, and place for the purpose of examining his knowledge in the action. It may, and often does, require that all data in the form of files, notes, photographs, plans, correspondence, etc., pertaining to the case be brought to the session and, in correct legal language, is call a subpoena duces tecum. Because a subpoena is a legal document, ignorance or refusal to adhere to its requirements may bring a fine, or in some instances, a brief stay as a guest in the local jail.

Usually, by the time an expert is served with a subpoena, he has already been involved in the case for a period of time. He will have his file completed - or nearly so –which contains information that will be the basis of the session. Upon receipt of the subpoena, his immediate response should be to call his attorney client reporting this fact. This puts the attorney on notice so that scheduling of appropriate dates and times may be arranged for a pre-deposition conference, and the deposition date can be adjusted to fit schedules. The pre-deposition conference is a meeting of the expert and his attorney to discuss the forthcoming deposition to review potential questions that will be asked, and to coach the expert on presentation of his answers. The expert should make certain that fees for his appearance at the deposition are paid by the deposing attorney, according to his fee schedule. Some deposing attorneys will try to skirt the fee situation by trying to pay lower "percipient witness" (non-expert) fees rather than those required by the expert. Assistance should be solicited from and assured by his attorney for timely payment at the pre-deposition conference.

The attorney and his experts are a team effort. The attorney will usually have an outline of a game plan for the forthcoming trial. The expert's job is to supply information for his client to pursue the case. One of the most important tasks for an expert is to determine the extent of his attorney's knowledge in the field of issue - and in particular, his understanding of how the processes work within that field. For example, if in construction, how well versed is he in building design, engineering and the methods and materials used in contracting. Until the expert has a good idea of his client's knowledge, much of his advice and information will not be understood. To gain this understanding, it is usually best to take a direct approach and query him on various aspects of the field and his knowledge of how things work in the process, including commonly used vocabulary terms. Upon gaining a grasp of his understanding, educational assistance can then be given to acquaint him with the specifics of

85

the field as pertains to the issue at hand. It is vitally important to know the attorney's game plan and the position the expert fits into the overall scheme. Once this is known, he can direct his information and advice to the proper area.

An understanding of the attorney's knowledge in the field of issue is usually assessed during the first meeting. It is a time of exchanging ideas and planning the course of action for the investigation. The subsequent predeposition conference is where the expert brings his findings, opinions, and conclusions for discussion. During this meeting, all details are worked over and smoothed out for presentation during the deposition. Timing usually calls this meeting to be with a few days or a week before the deposition.

The predeposition conference with the attorney, then, serves four purposes: (1) to learn his game plan for the case, (2) to expand his knowledge of the field pertinent to the issue as necessary for his complete understanding, (3) to supply him with the expert's thoughts, ideas, opinions, and conclusions before they are revealed to the opposing attorney, and (4) to allow him to coach his expert on his presentation and tell of expectations during deposition questioning.

Meeting with the attorney prior to the deposition also puts the expert on track as to what is expected of his testimony – the important items to be stressed and others to be avoided, if possible. The goal in a deposition is to reveal only enough information to give a good indication to the opposing side that they are faced with a knowledgeable and competent expert who has examined the facts of the case and can present them in a courtroom situation with poise and expertise.

Coaching the expert is a very important part of the overall plan. It serves the purpose of teaching him to control his actions during the deposition when responding to questions. The expert must keep in mind that the deposition process is one of discovery, and the opposing attorney's sole purpose is to ferret out information that may be helpful to his side and, if possible, damaging to the expert in the subsequent trial.

At deposition, the expert should be cordial, displaying his personality as a friendly and confident professional who is well prepared by having reviewed his file thoroughly and, if possible, made a recent visit to the site of the problem. He should be cooperative, avoid volunteering information not specifically requested, and avoid revealing any item that might prove embarrassing to either himself or his attorney. The expert should be able to answer any reasonable question that would demonstrate expertise in his field. Again, one of the opposing attorney's prime goals is to discredit the expert – and he will go after it should the opportunity present itself.

During the questioning process, the expert should listen carefully to the questions asked. If he doesn't understand the question, he should request that it be read back or repeated until he has a thorough understanding of what is said. Some questions may not be proper to the case and, therefore, lack foundation for answering. The expert's attorney will usually object to such questioning and advise the witness that he need not answer. If the expert needs to consult with his attorney regarding a question, he has that right and may request a brief recess.

The best tactic for keeping control of one's thoughts during questioning is to avoid quick responses. One should hesitate for a few seconds to digest the question and form the answer; this also will give his attorney the opportunity to object or make necessary remarks for the record pertaining to the question.

Thus, the objective of an expert during deposition is to answer questions to the best of his knowledge in the fewest number of words and avoid volunteering additional information.

The answers should be true to the best of his knowledge; no answers should be given that are fabricated to shade the truth. If the expert does not know the answer to a question, he should so state. He also should avoid giving answers to questions in which his memory is not clear; he should not guess, but state that he doesn't remember. While the expert's goal is to objectively serve and assist his client, the facts that he has uncovered and the conclusions drawn therefrom must stand by themselves regardless of the direction they may take. It is the attorney's position as to what to do with the facts as they exist.

The expert witness must be able to make his points during the questioning process as outlined with his attorney in the pre-deposition meeting. To do this, he must follow the game plan and make his points as the opportunities arise. He should also maintain his composure even when under attack by the opposing attorney. In the heat of the battle, he must clearly state his findings and opinions without losing composure. If he becomes rattled and confused, he is of little use in presenting his thoughts clearly so that others may understand his view. One of the many tactics used by attorneys in the trial cross examination process is to try to beat down the witness in any manner possible to make it appear that his testimony has no value or credibility.

Attorneys come in many types and personalities. During deposition, most are usually very friendly, respectful, and polite. They want to appear as reasonable fellows who are just trying to do a job. Their purpose is to cause the witness to drop his guard, thereby responding openly and freely to their questions. Most would like the atmosphere to become just a friendly chat about the situation. In reality, they are plying a tactic to gain information that one would not normally reveal in a more adverse climate.

Depositions normally follow three areas of questioning after positive identification of the witness is made: (1) education and training, (2) professional work experience, and (3) opinions, conclusions, and recommendations. The first two categories are searched to determine if the witness is truly an expert in his field and his background is indeed applicable to the issues in the case. The third category is where most of the heavy action occurs.

In the first category, questioning allows the attorney to establish the expert's formal education and/or technical training through schooling and skill-learning courses. It will normally include all earned certifications, degrees, licenses, etc., with a naming of technical schools, trade schools, colleges, universities, correspondence courses, government sponsored skill training programs, and the like. The times attended and the dates of obtaining certifications and licenses also will be of importance in establishing a track record. Continuing education in the expert's field of work will be questioned as well as any special training programs attended by the expert offered by his professional or trade associations.

Once the formal education and training portion have been established, the questioning then delves into the work experience history of the expert. This will include a chronological listing of all positions held throughout his career, including his advancement in the field because of to increased skills leading to his present position and degree of responsibility. Additionally, it may include articles and books written, formal courses taught, honors received, membership in professional and trade associations in his field, and positions of responsibility held.

One convenient way to shorten this rather lengthy questioning process is to supply the deposing attorneys with a personal resume' of his academic/training background and work history. This information can be contained on two or three typewritten pages and presented to the opposition at the proper time. It is known as the Curriculum Vitae (C.V.). It adds a touch of professionalism and also is of great assistance to the witness as it generally avoids a tedious search of his memory about his prior history. It is important to make it a thorough, but brief, representation. There will be questions derived from its contents, especially as to how his history applies to the case. It also is wise to make a quick review of this information just prior to the deposition.

The final category is the area where the remainder of the deposition will be concentrated. Once the attorneys are satisfied that the witness appears to have the appropriate academic and work experience qualifications, they will then concentrate their efforts on learning what the witness knows about the issue. This usually begins with a detailed search for the facts and the avenue in which the expert approached the task of forming his opinions and conclusions. The usual procedure is to go through the witness' complete file item by item questioning in detail every piece of information contained therein. This may be a very lengthy process, depending upon the importance of the witness and his position in the case.

To illustrate an example of a typical deposition, the process begins with a visit from a nondescript person who enters your office and asks to see you. As you come forth to greet him, and he's sure you are who you are, he hands over a piece of paper, thanks you and leaves. The man who just departed is a Process Server – and they can be very cagey in their approach to avoid the rapid disappearance of the person being served. Upon receipt and a reading of the paper, it reveals itself as a subpoena. It requires your presence at a given place, time and date and usually requires the accompaniment of your complete file on the case.

The next step is to call the attorney whom you are assisting on the case and inform him of the subpoena. He will then work out a time and date with the deposing attorney if the one designated is not available for you. He then sets up a predeposition conference with you to go over the case, which is usually within a few days prior to the deposition.

At the conference, he reviews the case and discusses your role as his expert. This includes going over each important point in your report that will accompany you to the deposition and reveals the names of the opposition's experts; this gives the opportunity to gain some insight into their expertise. He also will give guidelines on answering questions and will present his expectations of the opposition's method of examination. If he has previously given you depositions of the opposing expert's testimony, he will ask your opinion on their findings. They are usually quite lengthy documents. To have a quick recap of the information they contain, clerks and para-legals are used to summarize the depositions into two or three pages. It has been my experience, when reviewing these summaries after reading the depositions, that they often miss valuable information or have recorded it incorrectly. Therefore, I always read the depositions highlighting pertinent areas for reference; I do not review nor rely on summaries. Additionally, if it is at all possible, return to the scene of your investigation at this time to refresh your memory. Even though several years may have passed since the last visit and changes may have occurred, just viewing the scene again will recall thoughts forgotten. Photographs are good but not a substitute for live viewing.

Depositions may run as short as a few minutes or as long as several days, depending upon the complexities of the case and the extent of involvement of the expert. They are usually scheduled to take place just prior to trial so that all information gathered is fresh in the minds of the participants – usually within six to eight weeks of the trial date. Upon timely arrival at the designated place and date of the deposition, you should be well prepared both physically and mentally for the questioning period to follow. While the deposition process is less formal than being in a court of law, it does represent a legal proceeding and is not to be taken lightly. In attendance will be the deposing attorney, your attorney, plus one or more other attorneys who will also ask questions for the record. Also present will be the Court Reporter. It is important to establish just who the attorneys are in this gathering, whom they represent, and their interest in the proceedings. Knowing such, you will be better able to judge how your answers should be worded and information imparted.

The Court Reporter is a licensed shorthand professional who is hired by the deposing attorney to record the proceedings. All statements made during the deposition are recorded by a shorthand machine during the entire time of the deposition. When short breaks or lunch hours are taken, the record stops; it resumes again upon reconvening and the deposing attorney states that they are "back on the record." Any person attending the deposition may have any part of the recorded session repeated by the Reporter upon request. Another duty of the Reporter is to mark all physical evidence, such as reports, photographs, etc., with exhibit identification numbers when requested and include such information in the record. Your entire file may also be copied at this time by the deposing attorney's staff. A very important point to remember is not to testify during the time while your file is away from you being reproduced; you must always have it available to you for reference when answering questions. Don't rely on memory. If pressed for answers to questions without your file, simply state that you will have to look the information up after the file is returned.

Being a legal process, it begins with the Court Reporter swearing you in as in a normal court trial. Upon completion of the oath, the deposing attorney, who is the lead person for the opposition, will open the process by asking for your full name plus the spelling of the last name for the record. He will continue by explaining that all statements made henceforth will be recorded and then briefly explain the deposition process ending with the question of your understanding of the proceedings.

At this time, it is well to be reminded that the purpose of the deposition is painfully obvious; the opposing attorney's desire is to know to the fullest extent your history, background, thoughts, opinions, conclusions, and recommendations on the case before trial. It will give them a good idea of how strong or weak their case may be in light of this information, thereby giving an indication of whether to settle or push on to the trial itself. Additionally, the deposition allows the opposing attorney's experts an opportunity to review your thinking in the matter. By putting you through a very thorough questioning, it may expose flaws or errors that will create embarrassment or loss of credibility for you during the cross- examination phase of the trial.

Therefore, as an expert, you should always be aware that you are in the enemy's camp. You should be thoroughly prepared for the encounter with your testimony rehearsed, your demeanor one of calm and poise, your actions one of thoughtful consideration, and your guard up. Never drop your guard until safely away from all participants of the proceedings, except your own attorney.

Keep your answers to questions as short as possible – most should be answered with a simple "yes" or "no." Elaborate explanations should be avoided; only enough information should be given to make your points, and remember to pause a few seconds before answering. All during the proceedings, your attorney will be making objections to questions asked, so the pause is essential before answering. Another important item to remember is to speak at a normal pace. The Court Reporter has to record all conversation during the proceedings, and speaking too rapidly will require you to restate your answers for the record. This can prove to be both embarrassing and shows a lack of composure – not to mention the mistakes that the Reporter can make in trying to understand what is said.

As the deposing attorney begins querying your academic background, it is prudent to have a copy of your C.V. in front of you to assist with your answers. It is a good idea to have important dates marked in the margins of the pages to respond correctly to the questions that will surely be asked. Such questions will be the dates of your licensing or certification – that will give him a good idea of the length of time that you have been engaged as a professional. If you have taken special courses of study or taught formal classes, such dates should be readily available. The purpose of the inquiry is to ascertain that your academic/training background is indeed correct and timely to acquire specialized knowledge in your field.

Another piece of information that you should have immediately available is your history of testifying as an expert. This is a choice field of mining for deposing attorneys. Many folks ply their trade as "experts" on a full-time basis. Most are older people with a lifetime of experience in their respective fields and enjoy the opportunity to keep active in the business and make use of their large storehouse of knowledge and experience. However, there also are those who are "professional experts" who have good academic credentials, but relatively little actual work experience. It isn't for me to judge what is right or wrong, good or bad, but the general feeling that I have gathered from the legal community is that such "experts" are not true professionals but something less and, therefore, cast a pall upon their testimony. One question that is always asked is "what percentage of your work is as an expert in your total practice?" The best answer seems to be about 10% expert and 90% regular practice – and even a 50% split appears to be grudgingly acceptable. But with the exception of the older person, one who is a 100% full-time expert is thought of as a professional hired gun; a point that will be made very clearly in court to the jury.

Another of the main items of interest to the opposing attorney is possible bias on the part of the expert. Over zealousness to assist his attorney client may lead to the suspicion that his findings and judgement are not impartial and that he is merely a "yes man." Additionally, if the expert regularly serves his attorney client, bias may be claimed also. These suspicions may give the opposing attorney the welcome opportunity to make the claim at trial to discredit or even impeach the expert. To effect credibility in court and serve his client to the best of his ability, he must maintain a position of neutrality stating his unbiased findings, conclusions, and recommendations based upon competent and reliable material gathered in the case.

In any event, it is necessary to have your legal experience as an expert documented for immediate response. You should know approximately how many investigations you have performed, the types, the number of depositions you have given, and how many times you have testified as an expert at trial. It is also important to be able to put your experience as an expert into a time perspective; how long have you been serving as an expert and when were the dates of prior cases. The inclusion of a list with this information will readily reveal a thoroughly prepared file. Obviously, this gives the deposing attorney an idea of whether you are a veteran expert or a neophyte – and how to proceed accordingly.

As he moves on to your work history, you will be asked to elaborate on the projects under your direction; specifically, if there are types that are similar to the one under consideration. For example, if you are testifying as an expert on wood frame home construction, he may ask how many houses you have designed or built – and how many are in the same class or price range as the one at issue. His object is to see if you really do have such experience or just a general knowledge of the situation. Thus, if you are an expert in the field of wood home construction and the issue is regarding a house made of concrete

95

block, he will attack your knowledge and expertise on the use of concrete block in house construction.

Another example was a case on which I was recently deposed. My C.V. indicated that among the many buildings that I have designed were some 423 houses. I was queried on the number of different types of models, the materials of construction for the different types, how many were one story or two story, how many were single family residences, how many were condominium units, how many were of wood floors, how many were concrete floor slabs on grade, and so on. No one can remember such figures, only general percentages could be given as to the different categories. Nevertheless, such questions need to be answered to the best of your ability in a calm, thoughtful, and reflective manner. Such detail is approached by some deposing attorneys. My reaction was that it was a ploy to try to prod me into losing my composure. Simply answer the questions as best you can. My final response was that such information was contained in my office files and would be made available, if desired.

The main source of the questioning will be directed at the contents of your file. It is for this reason, as addressed in an earlier chapter, that all unnecessary papers, notes, etc., be discarded. Notes often contain information that can later be found in error, obsolete, or not pertinent to the issues; as such, it may be another source of questioning to attempt to discredit you as an expert. The file should contain only information that is essential to the case: your pertinent notes, reports, photographs, calculations, sketches, plans, a log of all inspection and conference dates including attendees, etc., which will be used by you to establish and justify your opinions and recommendations. Also, included should be copies of billings to your client, including hours spent and fee rates for the case. The file should be in a neat, concise package with all documents arranged in order of time – from the beginning of the investigation to the end of the work. In some cases, where there are many documents, an index to the papers should be made and attached to the inside cover of the file, a touch of professionalism and a real necessity for finding information quickly.

Depositions usually are rather tedious for the expert; he may be under constant questioning for many hours and has to try to remain sharp. Fatigue is an important factor in the process. As one grows tired and somewhat restless, he has a tendency to lower his guard by unconsciously rushing the answers. This also is a timely ploy often used by the deposing attorney to elicit answers that would have been more tactfully given earlier in the deposition during a less tiring time. This is also usually the time when the deposing attorney begins to use his heavy artillery – when you are tired and want the session to conclude. The best method to prepare for this is: (1) get a good night's sleep before the deposition, and (2) request a break at least once every two hours for a brief rest period.

When the deposing attorney states that he has completed his questioning, the remaining attorneys then proceed with their questions, one at a time. When the last has finished, the deposing attorney – and usually some of the others – state that they have just a few more questions. This back and forth game goes on until all questions are answered. The source of the original questioning comes from your file; subsequent questioning derives from your answers to previous questions asked by any one of the opposing attorneys. For this reason, it is again emphasized that you must keep your answers as short as possible, preferably with a simple yes or no. When an explanation is necessary, keep it short and simple. The more you talk, the more questions can be engendered from the new information. As noted in the beginning, the deposition process is one of discovery to obtain information that will aid the opposition's side of the case.

Though depositions vary in type and length, they are all similar. Content depends upon the type and style of the questioning attorneys. Various tactics are employed to elicit information from the expert. One in particular is the "hypothetical case" type. The attorney will create a fictitious situation very similar to the one in dispute and then, after assuring the expert that it is "just hypothetical," asks questions to try to bring responses to that case that are similar to the conditions in the actual issue. By so doing, he tries to create a parallel of thinking that is an effort to bring forth answers that will try to change the expert's opinion; if successful, he will then make the expert explain that if such was true in the hypothetical case, then why it is not in the present issue. These are used to trap the expert into a different line of thinking thereby creating confusion in his mind and, therefore, in his findings and opinions in the real issue. The best defense for this line of questioning is to be alert and watch for the subtle differences between the two cases; then point them out and put it to rest by stating that the facts or bases are not the same. The expert has to be on his toes during the process.

Another ploy, which I have had pulled on me several times, is at the end of the deposition. Just when everyone is tired and packing up their papers to leave, the deposing attorney says that he has just one more minor question – then he hits the expert with his Sunday punch just when he has dropped his guard. Often the expert's response is a quick, off-the-cuff reply; it may be very damaging and often re-opens the whole discussion. The expert's guard should be kept up until the door closes behind him.

When all questions are finally put to rest, the proceedings are closed. The Court Reporter takes charge of all exhibits and the machine recording tape. At this time, the Reporter will state that you will be notified of the time and place where the deposition may be read and have the opportunity to make corrections of mistakes found in the transcript. Notice will be received via the mails of its availability. It is most important to review the transcribed deposition. It will contain incorrect words, terms, and even incorrect statements where the meaning was changed by an incorrectly interpreted word. You are allowed to read and correct the deposition, such correction notations are to be made on a form provided for that purpose.

This, of course, dramatically points out that you must use words in your responses that are generally known to the lay public; technical words should be avoided where possible. If you must use technical words and phrases, speak slowly so that the Reporter may carefully follow your voice while recording. Also, assist the Reporter with spelling by repeating difficult words, as necessary. Once you have reviewed and corrected the transcribed deposition, you will then sign your name indicating your approval with corrections noted. The deposition then becomes a legal document useable in a court of law; it is a tool of vital importance in the litigation process. Expert witnesses brought into the fray are important participants in determining the facts in a case. The information they provide for their attorney clients form the basis for the presentations of the plaintiff and the defense.

An important point to note is one of confidentiality. The expert must keep his information and findings regarding the case to himself. He must not discuss the case with others, i.e. use it as the basis for lectures, talks, or general conversation outside of those involved in his side of the case until such time as it is no longer a matter of concern.

In summary, the process by which opposing sides gain knowledge of facts is through the discovery process. An important part of the process is the deposition. It is a duly recorded legal session at which each side has the opportunity to question expert witnesses at length from the information gleaned

Robert J. Crawford

from their files on the case. It is helpful in sizing up the opposition both in information content and the personal abilities of the experts. The finalized deposition is a legal document and may be used at trial in place of the expert's personal appearance - provided it has been signed by the expert - or his signature has been waived by the opposing attorneys, or after 30 days of official notification to the expert.

For the expert, a deposition is a tough, grueling process of questions and answers. He should be poised and confident in his demeanor, unbiased and factual in his information, and steadfast in his opinions and conclusions. He should also relate information only on the questions asked and not be given to lengthy explanations. He should be aware that the sole purpose of the deposition is to obtain all possible information from him to refute his opinions and conclusions and to discredit him as a credible witness. The deposition session is conducted in the enemy's camp and under the enemy's control. Survival is a matter of keeping control of one's emotions, remaining calm, and sticking to the facts as determined in the investigative process. As in any adverse situation, one must keep his guard up until out of the enemy's camp.

CHAPTER 7: PREPARING FOR TRIAL

Most of the problems addressed by the expert in the course of business will never reach the courtroom. Sometimes, though, the parties are at loggerheads and are willing to battle it out in the judicial arena. It has been my observation that when cases go to court, the only winners are the lawyers. Maybe satisfaction is something, but from a dollars and cents standpoint, it is a seemingly rare occurrence where the plaintiff actually gets what was sought.

The present trend in legal circles is to try to settle cases before going to a full court trial. There are several reasons for this. First, today's courts are filled to capacity with criminal cases that take precedent over civil actions causing long delays in obtaining a courtroom and a judge to hear the case. Secondly, settlement brings funds to the table without the significant legal expenses involved for attorney's fees at trial. This is why many defendants, such as insurance companies, make their first move toward settlement knowing that the cost will be less than if they go to trial. Their reasoning is that control can be placed on the final award as well as the attorney's fees. Trial brings the reality of not only significantly higher attorney's fees; but, if they lose, the award may be much higher than that which could be reached by settlement. A third reason is that the attorneys involved in the case are usually working on a percentage-of-award basis. Their desire is, hopefully, to collect on the settlement amount, thus avoiding the additional work required for trial – and the chance they may lose and receive nothing for their efforts.

Regardless of the reasons to proceed on to trial or to settle, if the court option is selected, the expert plays an important part. He must be well prepared for his role as a competent and knowledgeable practitioner in his field of expertise, presenting himself both intellectually and emotionally capable of stating his opinions in a clear, calm manner during both direct and cross examinations. He also must be prepared to act as a coach to his attorney client by assisting in preparing questions directed to the opposition's experts during cross-examination. Lawyers, for the most part, are adept at courtroom dramatics, but they cannot know everything. A good expert will educate them on the questions they should be asking and the meaning of the responding replies. In preparation then, the expert's job is twofold: he must prepare himself for the tasks ahead, and he also must work as a coach to his attorney for querying the opposition's experts.

PART 1: PREPARING FOR TRIAL

- A. TIMELINESS
- B. SITE REVISIT
- C. EXPERT'S FILE
- D. DEPOSITIONS AND REPORTS
- E. OFFICIAL DOCUMENTS, CODES, ETC.
- F. STUDY/THINKING TIME
- G. READING THE OPPOSITION
- H. COURTROOM PROCEDURE

A. TIMELINESS. When the date for the judicial proceeding is approaching, the attorney, in working with the expert, will make contact setting a date for a strategy meeting. This usually will be about four to six weeks prior to the trial. His intent will be to review the case in general and the expert's findings, to query the expert on specific items, and to explain in general his approach to be presented in court. He will also discuss information discovered during depositions given by the opponent's experts. In the present state of litigation, this meeting takes place approximately two years after the expert has completed his investigation and forwarded his report to the attorney. A lot of water has passed under the bridge since, memories have faded, and conditions of subsequent cases with similarity become confused with the one under consideration.

As an example, I was retained to determine the cause of distress in the roof framing systems of several relatively new residences in a subdivision. After completing the investigation, I prepared my report as to the cause with recommendations for repair. This particular building style was a popular model used by many subdivision developers throughout the locale because of its attractiveness and functionality – a best seller of the models offered. My investigation discovered each building of this model suffered from the same defects.

Sometime later, another attorney engaged my services to investigate some roof problems in essentially the same models in another subdivision. My investigation found the problems to be similar, but not identical, to the first study. When the original case came to trial some two years later, I had the dilemma of trying to recall the exact construction of the earlier study from the later one because of the similarity of the buildings. My memory was confused in trying to remember specifically the differences between them. Fortunately, my file of photographs, notes and sketches, etc., of the original enabled me to recall precisely what was needed; without it I would have been unable to correctly separate the two studies.

This is precisely the reason why it is imperative that the expert's file, usually generated some two years prior, needs to be complete to bring all facts of the case back to life as if they were gathered yesterday. The expert who does not undertake a complete and thorough study of his file immediately prior to trial is doing himself and his attorney client a disservice.

B. SITE REVISIT. If at all possible, a new visit to the site where the problem occurred must be made. Obviously, things probably will have changed during the interim, but the opportunity to once again review the site will bring back memories of the original investigation. Of course, photographs taken during the investigation will be of great assistance, but a comparison of them with what is seen on a current visit will greatly enhance recall of the conditions that existed at the prior time. This will afford the opportunity to see what changes, if any, have been made and to allow for current comments on what was seen.

As an example, I made an investigation of a problem in a rather complicated and archaic roof-framing system of an old building. My photographs taken at the time essentially covered the entire structural roof system. Because of the congestion in the attic area, the photos did not clearly show a particular portion of the framing that subsequently became an issue. Because the investigation had been conducted several years prior, my memory was somewhat hazy concerning the specific location of the problem area. I was aware that I would be queried about it, wanted to refresh my memory as well as to review the problem, and render my opinion accordingly. I was fortunate enough to regain access to study the area of the new concern. The revisit could, of course, have been denied, especially if the site had been controlled by the opposition. Had it been denied, I would have had to rely on the photographs and memory from two years prior for testifying in court. In a denial of re-access, photographs are of inestimable value in refreshing the memory of the original conditions at the time of the investigation.

It should be pointed out that the expert who represents those controlling the site usually enjoys relatively unrestricted free access for discovery. The opposing experts are generally not so fortunate and usually are limited to only one visit. Newly discovered information usually affords the opportunity for opposing experts to revisit a site; however, often the opportunity is granted with too short of a notice to respond or just plain denied. Reiterating, if access is available, it is wise to revisit the site to refresh the memory. Its presence will recall long forgotten thoughts to support your presentation as well as to show both your attorney client and the opposition that you are thorough in your work.

C. THE EXPERT'S FILE. The file generated during the investigation process represents the expert in the form of written documents. It is the basis of his involvement representing his investigation, findings, opinions, and recommendations in the case. It is, therefore, vitally important to see that it is complete, containing all information from the investigation leading to his final conclusions, as contained in the report.

Opposing attorneys like to see an expert's complete file. They take a dim view of editing by the expert prior to their perusal. The reason is that there may be some information that will assist their side in querying the expert during the discovery process. As anyone who has been through the information gathering process knows, the file is a collection basket for all data found during the quest. This includes all items that pertain to the issue; some are updated because of new findings, others become obsolete and no longer apply. Items that no longer have significance to the case should be discarded.

The point here is not to hide information from discovery, but to prevent unnecessary questioning by the opposition in matters that are not relevant to the investigation. As noted in Chapter 6, attorneys who are not conversant with the expert's field are on a hunting expedition. They generally do not have the technical knowledge to determine the value of the material that they are reviewing, and most just plow onward in hopes of finding something of note on which to further query the expert.

Inventory of the contents of the completed file should be edited to contain only the information pertinent to the case. This includes: names of participants, correspondence, all notes, sketches, photographs, plans, documents, depositions, interview comments of witnesses during the investigation, the completed report along with any appendage reports from others of assistance, and billings to the expert's client.

As the file is edited for accuracy and clarity, all data therein should be thoroughly reviewed to bring the information clearly back to mind. This activity is extremely valuable, enabling the expert to converse accurately on all findings, opinions, and recommendations of the case.

D. DEPOSITIONS AND REPORTS. Those portions of information that are not actually contained within the file itself, such as plans, reports, depositions, and items collected from various information sources, are also considered to be a part of the total file. These items also must be reviewed during the memory- refreshing process and referred to while giving testimony during both direct examination and cross-examination.

Depositions of experts from both sides involved in the case are supplied by the expert's attorney. A study of testimony given by the opposition's experts is very valuable as it reveals their opinions and recommendations on the issues of the case. When this information is known, the expert can discuss with his attorney his opinion of their testimony, pointing out weaknesses, inconsistencies – as well as strengths - on the other side of the case.

From the standpoint of being an expert, the most important deposition to be reviewed is of the expert himself. It is vitally important to refresh the memory on what was said during his deposition; this is the information the opposing attorney will use to try to trap the expert by his own words on a given subject or comment made during the deposition. It must be remembered, the opposing attorney's goal is to discredit or impeach the expert. Not thoroughly studying his own deposition sets him up for relying on memory that is several weeks or months old; a dangerous practice when in the hands of the opposing attorney.

E. OFFICIAL DOCUMENTS, CODES, ETC. The expert should thoroughly review all official references that supply the basis of his testimony. The rules, as published in various ordinances, codes, reference texts, etc., are usually at the heart of a case and must be clearly in mind to substantiate the opinions put forth. Every publication used in preparing the report should be readily accessible when appearing in deposition and court; pages should be tabbed and sentences highlighted for quick reference.

F. STUDY/THINKING TIME. After the expert has brought himself current with a thorough review of his file, hopefully a revisit to the site, a thorough reading of all depositions - with special emphasis on his own, and a study of the applicable official documents used in the preparation of the report, he should take a step back and contemplate on the overall perspective of the case. He should see his position as a team player and make sure that he has covered all items as requested in the original scope of work. Have all loose ends been neatly tied together without any unresolved issues? Do all the parts of the puzzle fit together neatly and securely? Has anything obvious been overlooked in the process? This is a general reflection on the process as a whole and how his efforts to provide information fit into the overall picture.

G. READING THE OPPOSITION. Once the expert has thoroughly reviewed his efforts in the case, he should turn to evaluate the opposition's experts. Having read their depositions enables him to gain insight into their value. Depositions reveal much more than the thoughts of the deposed. They also reveal their educational background, technical expertise, personality through the recorded spoken word, composure during questioning, how they approached their assignment, and how thorough they have been in carrying out their task. Much can also be gained from reading the expert's own deposition; it is surprising to discover that there is a difference between how one thinks he's presenting his thoughts and the way he appears to others. It is truly an educational learning process on presentation. In any event, by studying the opposition, the expert can judge how he measures up to them, or they to him, in the contest.

A study of opposing experts can be quite an educational process. Often the reputation of a given expert precedes his appearance, especially if held in high regard by his contemporaries and/or the legal community. I have been involved in cases where opposing experts were accorded extremely high standing in their fields. Perhaps they were to their peers, but in the arena of battle, they proved to be quite less than expected. They relied primarily on their reputation and "authoritative word" to carry the day. My observations have found that those who do the necessary "homework" on a case by being extremely thorough in the various phases of discovery are usually the ones who convince juries. So, it pays to do a little homework on your adversaries; it will certainly add valuable information to your plan of presentation.

H. COURTROOM PROCEDURE. All preparation undertaken by the expert leads to one specific performance – presentation in the courtroom. The presentation of information by the expert follows an established procedure. This will be taken up in detail in the next chapter, but the essence is the expert relating his information to the judge and jury through answering a series of questions asked by the attorneys. The format follows: (1) the presentation of the complaint by the plaintiff's attorney; (2) at the conclusion of his presentation, the defendant's attorney presents his case in defense. During the presentation of each, experts are questioned as to their findings first by their own attorney - called direct examination - and then by the opposing attorney -called cross-examination. This courtroom drama is played before those who will make the decision as to whether the plaintiff's side or the defendant's side is the most convincing.

PART 2: PREPARING COUNSEL

A   COUNSEL'S ATTITUDE AND PERSONALITY
B.   COUNSEL'S KNOWLEDGE OF FIELD
C.   COUNSEL'S STRATEGY/GAME PLAN
D.   OPPOSING COUNSEL'S STRATEGY/GAME PLAN
E.   MAJOR REVIEW/REHEARSAL
F.   COACHING THE ATTORNEY

A.   COUNSEL'S ATTITUDE & PERSONALITY. In order to be effective, the expert and the attorney have to be on the same wave length; they have to be able to communicate effectively with mutual respect for the exchange of ideas and information.   An attitude of friendly assistance has to be adopted; one that allows for ease of communication without a sense of social or educational restrictions. They have to work together. A case in point is an issue in which I was one of the team of experts in attendance; it was a complicated problem requiring many areas of specialized knowledge.   The attorney was a capable, confident individual with a pleasant attitude.   One of the experts was a college professor and an acknowledged leader in his area of expertise; unfortunately, he had an "ivory tower" attitude, that affects a number of teaching professionals. Outside of his academic circle, he proved to be a loner who couldn't adjust to being a team player. While his area of knowledge was valuable to the case, it was only one of the parts that contribute to the whole. His attitude estranged him from the group to the point where no one wanted to work with him. The attorney, while fully aware of the professor's wealth of knowledge, perceived the problem and, much to his credit, replaced him. From that point on, the attorney and his team of experts worked smoothly as a unit.

B. COUNSEL'S KNOWLEDGE OF FIELD. This brings up the extent of knowledge the attorney has in the field of expertise under study. Many attorneys have considerable knowledge and experience in the area under consideration - needing the assistance of technical expertise for the finishing touches to their presentation. Others are totally devoid of any knowledge of the area and require thorough tutoring by the expert. This is why it is so important in the initial conversation to query the attorney on his depth of knowledge in the field. Once his grasp is known, the educational process can begin. The purpose, of course, is so the attorney can make full use of his expert in extracting testimonial information from opposing experts and also be precisely knowledgeable of what he is presenting to those who sit in judgement. The educational process between the expert and the attorney is a continuing one with each educating the other; this begins at the initial meeting and continues until the case is judged at trial.

C. COUNSEL'S STRATEGY/GAME PLAN. When both the attorney and the expert are comfortable in the knowledge they share in the case, the attorney needs to outline his strategy for presenting the information to the court as well as his goals in the case. Obviously, the attorney wants to prevail, but often other factors in the issue are not known or apparent to the expert. That is why he needs to know and understand all issues in the case and how his input is used to effect the desired result. The expert, therefore, needs to know not only the goals of the attorney, but also his game plan for getting there. While experts normally are used for their expertise, often their understanding of the entire situation brings forth strategy, ideas, and suggestions that are a part of their knowledge as businessmen and participants in legal proceedings. Experts are usually widely knowledgeable in areas contingent to their own field.

D. OPPOSING COUNSEL'S STRATEGY/GAME PLAN. While it is highly desirable to know what the opposition has planned, some thought should be given by the expert as to the method of attack that is expected to come. This is generally a brainstorming session by the attorney and his expert. At this phase of the case, all information from both sides is generally known. The expert has stated his opinions during discovery, and he can expect opposing counsel to query extensively in that area. It is prudent, therefore, to anticipate in advance those questions that will be asked in order to be prepared with answers that are consistent with prior testimony. The expert can be extremely helpful to his attorney by originating a list of questions to be asked by him during the direct examination to bring out the major points of the case to the jury. A second list of questions should be made as well for querying the opposing experts during cross-examination.

E. MAJOR REVIEW/REHEARSAL. After the expert and his attorney have smoothed out the educational process and produced the lists of questions to be addressed to the experts, now is the time for a complete review of all major points in the case. A rehearsal should be conducted to ensure that the expert knows and understands all questions that will be asked of him during the direct examination with the answers they want the jury to hear. Additionally, they need to rehearse for what is to be expected of the opposition during the cross-examination period. This review also will bring up any new areas that may have been overlooked during the brainstorming process. As an example, during the final phases of review with a client,

I found newly published research data that had an application to the issue. This new information provided further foundation for the findings I was putting forth, thereby giving another reliable source for credibility.

F. COACHING THE ATTORNEY. No matter the extent of information the expert has related nor how bright and retentive the attorney may be, it is impossible for him to grasp and remember all of it. The main item of importance is that he understands the logical progression of the information in a step-by-step manner – as one foot follows the other. It is a coach/player situation wherein the expert teaches his craft, as it applies to the specifics of the case, so that the attorney may readily follow the meaning and applicability of the information presented.

Another very important and effective way to assist is being in attendance during the deposing of opposing experts. While it is not permitted at trial, the discovery process allows the deposing attorney to be accompanied by his expert for guidance in his questioning. I attended a deposition acting as a coach to my attorney client in one rather large case. The direction of the questioning and resultant responses had taken him deeply into a technical area, certainly beyond his limited understanding. At this point, in a quiet, aside conversation, I was able to explain what the responses meant and what further questions should be asked to get to the heart of the matter. After the session was completed and on returning to his office, I further enlightened him of the meaning of the deposed expert's statements so that he would fully understand how to use this information. Working as a team, the effort succeeded as he was able to discover important information that would have been missed if he had been alone.

Robert J. Crawford

In summary, the expert has two very important tasks in preparing for the legal contest. He must first thoroughly review his complete file with accompanying documentation, depositions, and reports to bring back all thoughts that supported his findings, analysis, conclusions, and recommendations. He must be ready to stand up to rigorous questioning by the opposition in their attempts to negate his testimony and/or impeachment as an expert witness. Secondly, he must fully educate his attorney-client in his field of expertise as concerns the issue under consideration. Further, he must be assured that his attorney fully understands what his findings and conclusions mean and how to put them to best usage. He must assist the attorney in formulating questions to be asked at trial both for himself to bring out his main points and to the opposition for digging into their findings and opinions and, hopefully, to cast doubt as to their validity and credibility. It is a challenging task and requires diligent work to be well prepared for the contest that lies ahead.

CHAPTER 8: THE JUDICIAL ARENA

All preparation efforts put forth by attorneys and their experts on both sides of an issue are directed toward one final production to be performed in the judicial arena - of which there are several forms. Most commonly known is the Court where formal trials are held and issues decided. On the second tier of the judiciary process is the Arbitration – which is less formal than the court trial, and it may take one of several forms. Another is called Mediation, the least formal and often the starting point of the judicial process. There are other variations of lesser popularity, but all fall under the auspices of the Court in which the lawsuit was filed.

It is in these areas of judicial governing that civil disputes are heard, judged, and decisions rendered. Decisions are determined by the merits of the presentations of the two sides in the case. Those pleading for the plaintiff, if they prevail, enjoy restitution for alleged damages suffered - often resulting in monetary awards. Conversely, those defending against the charges, should they prevail, are exonerated and thereby relieved of any further duty in the matter. Often the outcome may be somewhere in between these two extremes.

The process normally begins with members representing opposing sides presenting their information according to a prescribed procedure. At this point, the attorneys have originated, polished, and rehearsed their game plans with their experts. All participants during the process become performers with each following the precisely rehearsed role. The goal of each team member is to accurately and convincingly present information to strengthen its side of the case. Those who sit in judgement will render their decision based upon the information presented by participants from each side.

The expert witness may find himself participating in any of the various arenas of judgement. It is an established fact that lawsuits are extremely expensive, and, for this reason, the expert should have a general working knowledge of how each is conducted. While there are many fine points as to the functions and limitations of each, the purpose here is only to give an overview of their working process and what the expert may expect if called to testify. For precise legal composition and procedures of the various judicial arenas, consult appropriate legal texts.

JUDICIAL STRUCTURE

MEDIATIONS. Because of the preponderance of litigation in civil cases, most Courts desire, and many require, parties to a lawsuit to first bring their issue to Mediation. Mediations have the advantages of being less formal, speedier, more convenient, and can be held in an office meeting room. The process is initiated with each side agreeing on the selection of one person to serve as the Mediator, usually an attorney. The Mediator's job is to try to bring the two sides into agreement on a common ground of acceptance - a give and take on the issues - in an effort to resolve the dispute. There are no others sitting in the responsible position of impartial assistance than the Mediator; he is compensated for his time.

The process is similar to courtroom procedure where the plaintiff's counsel presents his case along with accompanying expert testimony to reinforce his claims. Defendant's counsel has the opportunity to query the plaintiff's experts during the presentation. Upon completion, the defense counsel follows the same procedure in presenting his side of the issues accompanied by the testimony of his experts. Again, plaintiff's counsel has the opportunity for cross-examination of defense's experts. During the presentation of information by the attorneys and their experts, the Mediator has the opportunity to ask questions for clarification, etc., for his further understanding. There is no Court Reporter taking minutes at the proceeding, nor are the experts sworn under oath.

Again, the purpose and efforts of the Mediator is to try to bring the parties into some sort of agreement to settle the dispute. The fairness of the settlement is entirely up to the parties to the Mediation. The Mediator has to remain impartial, does not pass judgement on the merits of either side's presentation but only tries to effect a compromise between the parties. Failure to reach a compromise paves the way for higher judicial action.

ARBITRATION. There are essentially two types of Arbitration: Binding and Non-Binding. When opposing parties select Arbitration for settlement of a dispute, they have the choice of making the final decision either binding - or not binding. It is a choice that is made prior to the Arbitration proceedings. In Binding Arbitration, the participants will be held to the final decision of the Arbitrator, the same as if the case were before a court of law. In Non-Binding Arbitration, the parties are not obligated to accept the judgement of the Arbitrator and may pursue further restitution or relief through the Courts.

Similar to the Mediation, Arbitration also is a less formal, quicker method to settle disputes. It can be held in a meeting room, usually at a lawyer's office, and discussions can be made across the room on points of testimony. The cost is less than a court trial and slightly more expensive than the Mediation. The Arbitrators are compensated, usually according to a schedule espoused by the American Arbitration Association. Arbitrators are usually lawyers, practicing or retired, or in many cases, experienced professionals in the field of expertise related to the issue at hand.

In Arbitration, there may be one Arbitrator who sits in judgement of the issues, or there may be a panel of Arbitrators who will determine judgement. In either type, the Arbitrator(s) is selected and agreed upon by both sides to the case. Where there is more than one, usually three is the popular number; each side selects one Arbitrator to represent his interests. These two, in concert, then select a third Arbitrator to act as a referee who is independent of any involvement and acceptable to each of the parties of the lawsuit. The procedure is similar to the courtroom process where the plaintiff's side presents his case along with testimony from his expert witnesses. This is followed by cross-examination of the witnesses by the defense counsel. On completion, the defense then makes his presentation along with his expert witness' testimony followed by cross-examination by the plaintiff's counsel. The Arbitrator(s) has the opportunity to ask questions for clarification for further understanding from anyone offering testimony. There is no Court Reporter taking minutes during the proceedings and experts are not sworn under oath.

After the proceedings are concluded, the Arbitrator(s) adjourns for review and deliberation. In the case of a sole Arbitrator, he will then refer to his notes of the proceedings, any pertinent documents, photographs, etc., for study. At the completion of his study, he will write a statement of his findings delineating his judgement in the matter. Where there are three or more Arbitrators, they will usually follow the same procedure. When they have come to their conclusions, they will meet and discuss their findings. Judgement will be on the basis of the findings of the majority of the Arbitrators. Again, Binding Arbitration settles the dispute with its judgement. Non-Binding Arbitration opens the door to further litigation - if desired by the party who feels he has a better chance in Court.

COURTS. When all else fails, when issues are not settled in Mediations or the various forms of Arbitration, the remaining path leads to the courtroom. Here is found the full formal procedure of the law. The case is heard in a very formal setting with strict adherence to established procedure and legal etiquette. The following presents a simplified overview of the procedure; in actuality, court cases may be quite complex. However, the point here is to acquaint the expert on how the process works.

The Courtroom is presided over by the Judge of the Court. He is the top legal officer of the Court and is responsible for the fair and just administration of the proceedings therein. He is referred to as "Your Honor" in all addresses within the courtroom, a distinction conferred upon him with the office. His position is to assure that all proceedings within his court are in accordance with the law and he rules on whether evidence presented is admissible as well as if expert witnesses are indeed qualified to testify. He sees that trial procedure is conducted within the letter of the law, makes judgements on issues brought before him during the course of trial, and maintains discipline in the Courtroom with the aid of his Bailiff, a uniformed police officer assigned to the court. In short, he is the person in charge of the courtroom and all that goes on within it.

Another permanent members of the court is the Court Reporter/Clerk, who duly records all spoken words uttered during the proceedings, keeps track of exhibits, and also administers the Oath of truthful testimony to witnesses. There is also the Bailiff, of course, and his subordinate officers, if necessary. During trial, the other Officers of the Court are the attorneys for the plaintiff and defense; each may have several assistants.

The Courtroom is divided into two parts, the judicial area and the spectator section. Separating the two is a low wood railing across the room, sometimes call the "Bar" or "Rail." The Judge presides from his high desk, call the "Bench" centered against the back wall of the room; the witness chair is either at the left or right side of the Bench, the Court Reporter's/Clerk's desk is below and in the area in front of the Bench. The tables for the plaintiff and defense counsels are on each side of the aisle inside the judicial area next to the rail, plaintiff on the right and defense on the left. The jury box, containing twelve seats, is located against the left or right side wall between the attorney tables and the Bench. The Bailiff's station is at the side door adjacent to the jury box. From the courtroom entry to the wood rail is the spectator section divided in half by the center aisle leading to the judicial area.

There are two types of judicial proceedings in the Courtroom. One is termed a "Judicial Trial," meaning the judgement in the case is made by the Judge alone. The second is the "Jury Trial," where judgement is made by a jury of twelve jurors under the explicit directions from the Judge. The findings of the jury may be altered or amended by the Judge as presiding officer of the Court.

A jury is composed of twelve registered voter-citizens who have been called and selected to serve as a juror by their County of residence. Before selection, each prospective juror undergoes questioning, called voir dire, by the attorneys from each side and sometimes the Judge, to test their impartiality in the case. Those who "pass" the test are accepted as members of the jury for the case. The Judge will require the jury to elect a spokesman, called the "Foreman," to speak for the group. Usually, there are two alternate jurors selected to fill in for a juror who is unable to continue to serve because of health or removal by the Judge. The alternates attend all proceedings with the designated jurors.

Courtroom procedure is very formal, and a brief walk-through will give the uninitiated an overview of the process to know what to expect. It begins with the arrival of the Judge from his chambers, which are adjacent to the Bench. He then calls the court to order. At this time, all participants in the trial are present: the Court Reporter/Clerk, Bailiff, attorneys for the plaintiff and defense, and the jury, if a jury trial. Expert witnesses for each side are not allowed to sit in court; their presence is required only when they are to present their testimony. Continuing, the Judge then asks the attorney for the plaintiff to proceed. The attorney then makes his opening statement, outlining the claims on behalf of his client, his program for substantiation, and expected judgement. This is followed by the attorney for the defense making his opening statement, presenting a denial of the claims and his reasons for such denial.

121

At his conclusion, the attorney for the plaintiff begins his presentation for substantiation of the claims by a calling a series of witnesses. The first is usually the plaintiff who will state the history of problems and the reasons why he wants relief from the defendant. To substantiate the claims, the attorney will then call on his expert witnesses. After their acceptance as duly qualified experts by the Judge and opposing counsel, the attorney then builds his case by laying a foundation of facts, findings, and opinions from his experts substantiating the claims. This questioning of the plaintiff's counsel of his own witnesses is called "direct examination." Exhibits supplied by his experts may be presented to the court to provide documented evidence of their testimony along with explanations of their meaning. When the attorney has concluded his direct examination of each of his experts, the opposing attorney is allowed to query each at the conclusion of their testimony; this process is called "cross-examination."

At the conclusion of the plaintiff's presentation, his side retires and the defendant's side takes the floor. The procedure is essentially the same for the presentation of evidence by the defense, their experts' testimonies, exhibits, etc., and concludes with the expert witness cross-examination process by the plaintiff's counsel. After all information has been presented, expert witnesses examined, testimony given, final arguments are then presented by the attorneys for each side, and plaintiff first, to substantiate their claims. Each concludes with a plea that their presentation of the case prevail. The Judge then concludes the proceedings.

If it is a judicial decision, one made by the Judge alone, he will then retire to ponder the evidence and eventually render his decision in the matter. If it is a trial by jury, the Judge will then instruct the jury to retire to the jury room, deliberate the case, and return when they have made a decision. At that time, and upon informing the Judge, the Judge recalls the jury to the jury box and asks the Foreman to present their finding. Upon delivery of the verdict, they are dismissed; the Judge pronounces judgement, and the case closed.

THE EXPERT IN THE JUDICIAL ARENA

The expert's appearance in each of the judicial arenas has one objective: to present his information to those who sit in judgement in a manner that they will clearly understand his testimony and opinions for the side of the issue he represents. To accomplish this, there are a number of considerations he must take into account. While the Mediation and Arbitration are less formal than a Court trial, the expert should present himself in the same manner – at his best - regardless of the informality. It is extremely important that he make a favorable impression on those who sit in judgement of the case. To gain a better understanding of those to whom he is to address, it is necessary to explore just who they are and the best way to reach them.

THE JUDGE, ARBITRATOR & MEDIATOR. Those who sit basically alone in judgement on an issue are usually well qualified to assess the facts presented by the opposing sides. When the outcome of a dispute between parties is to be determined by a judicial decision, the case is heard and decided by the presiding judge of the court. Judges are lawyers who have been placed on the bench by political appointment, usually the Governor of a State, and face re-election periodically. Most are well qualified to serve in the capacity and dispense justice fairly. Judges are human, though, and as often as not, they are quite taken with their authority.

Some are quite egotistical and demand deference to all within the confines of their courtroom. However, most are well qualified both mentally and emotionally to serve the system to the best of their ability. It is well to be on one's best behavior when on the witness stand as the Judge has total power over your presence. Besides, it just makes good sense to show oneself as a competent and respectful practitioner of his craft.

Arbitrators and Mediators also are usually attorneys, although some arbitrators are professionals in their fields of expertise, as noted above. While not on the same hierarchical plane as judges, they nevertheless preside over their legal proceedings in a manner similar to a Judge. Because the proceedings are less formal than court, the communication procedures are more relaxed. However, they are not to be taken lightly; their judgement is just as important as that of the court. It is, therefore, very prudent to be thoroughly prepared in all respects for each of the legal arenas in exactly the same manner.

JURIES. As often as not, the majority of legal actions that end up in court are decided by juries. As briefly outlined above, members of juries are selected from the citizenry by voter registration files. Those who serve on juries are generally lay people with little or no knowledge of the issues that come before them. They are generally from all walks of life, of average intelligence, and are not particularly happy about being called to sit in a jury box listening to a group of lawyers argue about an issue. True, some cases are more interesting than others, but my experience in the courtroom has observed that some jurors find themselves to be totally bored and uninterested – though usually trying to keep up with the proceedings. More than once have I seen a Judge rouse a juror from napping. It is to these people the expert must tell his story and tell it in such a manner that it will pique their interest and attention; in short, it must be both entertaining and convincing. Quite a tall order, but this is the nature of dealing with juries. They are the reason for all the pretrial preparation effort - right down to the final performance on the expert's day in court.

So how does an expert witness obtain and hold their attention? There are several ways this can be accomplished. To begin with, the expert has an advantage. When called into court by his counsel, he automatically becomes the center of attention. All eyes are on him as he makes his timely entrance, takes his seat in the witness box, and is sworn in by the Court Reporter/Clerk. The jury watches this newcomer in their midst, each making a visual appraisal of him and wondering what sort of a fellow this witness will be. So it is imperative that the expert makes as good an impression as possible, not only with the information he will offer, but also by his appearance and demeanor. He must be seen as what would be expected of a professional in his field: clean, neat, groomed, and dressed appropriately for the occasion, moving with confidence as he takes the stand. He should sit upright, wear a pleasant countenance appropriate for the situation, keep his attention focused on the matters at hand, and avoid looking around the room. After being sworn in, he should take a moment to arrange his files, etc., for quick, easy reference. Once settled in, he should direct his attention to counsel indicating that he is ready to begin.

At this point, the expert has made a favorable impression on the jury. He now has their attention.

From this time on, the jury will hear as well as see him. How he presents himself verbally will either continue the favorable impression – or become a detriment to him; it is all in his physical movements and how he speaks.

Once established, the expert needs to maintain his professional demeanor. His voice should be clear, steady, interesting, and loud enough to be heard throughout the room. It should not be given to abrupt changes in volume or pitch, nor delivered in a monotone. He should display a pleasing personality, keep a good grip on his emotions, and answer all questions calmly and directly. He should answer in short sentences - about 30 words or less – and avoid long, rambling, explanations where jurors will lose track of what is said. Where possible, information should be described in word pictures so that the jury can visualize what is said. Testimony given by the expert must avoid technical language. A little ingenuity should be employed to use words that any thirteen-year-old child would understand. He should make full use of examples to illustrate his points. One of the best is through the use of exhibits; these are items that the jury can actually see and gain the meaning of what is being offered.

As an example during testimony at trial, I had built a four-foot by four-foot panel that represented a portion of a wood-framed wall. It was covered on one side with gypsum wallboard nailed to the wood studs. On the other side, I also covered the studs with gypsum wallboard, but added a texture finish coating over the entire surface as would be found in a finished wall in a home. My point was to show the jury the difficulty in re-nailing the wallboard to the underlying wood studs - as shown on the unfinished side - to a finished wall because all wallboard joints and studs were hidden by the texture material. A simple example of an exhibit, but it was extremely easy to make my point using the exhibit; to verbally describe this difficulty - to try to get them to visualize the nailing problem - would not have been as convincing.

It is also imperative that questions answered by the expert be directed to the jury, not to the person asking the questions. It must be remembered, the jury is the body that will make the decision in the case, they are the ones who are the object of all efforts. It also is vitally important to have the expert's counsel determine who in the jury has the most knowledge or experience in the field of the expert's testimony. Those jurors then become the ones to whom the expert directs his attention – as if he is talking directly to them in presenting his answers. He should make eye contact with them and maintain it throughout his answers, returning to the questioner only when ready for the next query. Later, during deliberations, these jurors will educate the others as to the meanings of what was presented. Opposing attorneys will not like this tactic, but the expert is not there for their benefit. The jury is the key to the decision-making process; it is they who are to understand the information offered by the expert.

Summarizing the exchange between the expert witness and the jury, the expert must present himself as the most knowledgeable person present in the area of his expertise. He must do this in a professional manner offering information in short, easily understandable sentences, showing exhibits to further clarify difficult points, and presenting all information aimed directly at the jury. He must remain unruffled, pleasant, and interesting. He must also show that the obvious attacks

127

by opposing attorneys do not upset nor shake him from his opinions in the case. His job is to tell his story in the clearest and most convincing manner of which he is capable, and he tells this story to the jury.

EXPERT'S ROLE. It should be understood by now that the job of the expert is to get his message across to the jury. He does this in the physical appearance of his person and the verbal offerings of his testimony. The vehicle for this transfer of information is through examination of the expert.

Direct Examination. The opening of the direct examination by the expert's counsel will be to convince the presiding officer, judge, arbitrator, or mediator that the expert is qualified to testify in his field of expertise. This entails a brief questioning period to bring out the educational background and experience of the expert. On acceptance, the process continues with the expert's counsel following the previously planned and rehearsed questioning schedule. The expert's testimony will follow the guidelines that were designed to bring out all information he has to offer in the matter. While so doing, he will utilize his file for reference as he moves from point to point reinforcing his story with exhibits, such as models, diagrams, maps, charts, graphs, photographs, models, etc. It is imperative the he be straight forward and direct in presenting his opinions, avoiding any indication of doubt. He should have on hand all back-up information, such as official documents, codes, etc., readily available for supporting his opinions when challenged. Above all, he must never be, nor give the appearance of being, a pawn of his attorney; there is no quicker way to impeachment than to be less than honest. When his counsel has completed his questioning, the opposing counsel will then begin his querying.

Cross-Examination. This is the time when most expert witnesses become apprehensive. The opposing attorney's job is to try to discredit the expert's testimony in front of the jury. It is certainly no fun to be grilled before an audience by an aggressive attorney bent on discrediting the opposing expert's opinions. He will turn up the heat, try to rattle the expert, and try to shake loose his opinions by casting doubt based on information gathered by his own experts. The expert must resist these pressures as well as those brought by questions to which he doesn't know the answers. He must also keep control of anger by not allowing it to surface during attack. If he doesn't know the answer to a question, he should so state that "he doesn't know." He should never guess or speculate in answering questions. Not an easy time for the expert witness.

In answering questions under cross-examination, the expert should be honest in his replies and give direct answers to the queries. He should always be fair; if he honestly disagrees, he should so state and give the reasons why, if asked. Again, it is important not to become angry at impertinent questions or inferences; it is especially important not to accept a challenge to do battle – no matter how much of a temptation - as the jury is watching all of this and will begin to discount the credibility of the expert. It is of significant assistance if the expert can pre-think questions he might be asked during cross-examination. It saves a lot of scrambling and allows for calm replies. He also should be quick on his feet in thinking through his answers while also determining the direction the attorney is taking him. It is well to remember once again, the expert is the

129

knowledgeable person in his field in the courtroom. The opposing attorneys only know what they have been told by their experts, plus any additional reading in preparation for trail. They are very good at bluffing, giving the impression they know as much or more than the expert. The expert should always be aware of this fact and not back down from an attack that he can successfully defend.

The best defense is to remain calm, poised but not tense, listen carefully, and remember that he is the expert, not the attorney. His opinions have been carefully determined and, to the best of his knowledge, they are the truth of the matter. He should stick to his guns answering all questions to the best of his knowledge and keep consistent with his beliefs. He also should show that he is unbiased and willing to listen to other opinions presented. If some subsequent information is brought forth of which he was unaware that sheds new light on the subject, he should be willing to assess these new findings in light of his expertise and judgement. If they have relevancy, he should so state but also remind the opposing counsel - and those in observance - that this is the first he has known of this information. If it is not relevant, he should so state by discounting it as not pertinent to the issue giving his reasons. This should be followed with a reiteration of his own beliefs and opinions. The expert who changes his testimony, without just consideration during trial, is a candidate for impeachment by the opposing counsel.

The most important item of review for the cross-examination phase of testimony is a thorough study of the expert's own deposition. The only ammunition available to the opposing counsel are those statements made during deposition and from written reports. The opposing counsel's questioning will try to undermine them in an effort to force the expert to contradict himself. It is extremely difficult to remember all statements made during a deposition given several months past. If it was given several years ago, the accuracy of what was said is virtually lost from memory. The opposing counsel will try to confuse the expert as to his own statements in an effort to discredit him as a witness. It is a well-worn tactic, but very effective, if the expert has not thoroughly reviewed all statements made by him during the discovery phase of the case. This also is the reason why it is imperative that the expert review a copy of his deposition as soon as possible after it was taken. It affords him the opportunity to correct mistakes made by the Court Reporter in misinterpreting information spoken during the discovery examination. Again, the expert witness who does not thoroughly review his deposition, as well as all reports authored by him, is setting himself up for a difficult time on the witness stand, a loss of credibility with the jury, or a good possibility of being impeached.

In summary, the legal arena is where all efforts in a case are directed. The proceedings may take place in a Mediation, an Arbitration, or move on to the Court of Law. The general construction of each is similar as far as the expert witness' involvement is concerned. While the informality of the Mediation and Arbitration is less intense, they are just as important as if the case were heard at Court.

The expert's job is to clearly and effectively present his information to those sitting in judgement. He needs to package himself properly to be representative of the level of expertise he espouses; to be thoroughly prepared and rehearsed is the goal when entering the judicial arena. How well he accomplishes this will have a great effect on the outcome of his testimony. He can

Robert J. Crawford

fully expect to be under attack by opposing counsel to discredit him along with all information he brings to those sitting in judgement. He can survive if he is prepared and ready for the task before him. He needs only to be armed with the truth of his findings, backed up by his education, knowledge and judgement in exercising his opinions. Once armed, his task is to present this information to those sitting in judgement in a manner that clearly substantiates his professionalism and fairness.

CHAPTER 9: AFTER THE BALL

A professional in any field of endeavor will always analyze past performances. Just as the golf, football, or chess pro reflects on his recent competitive efforts, the object is to review his performance. For those things done correctly, he can pat himself on the back remembering to use them again in the future. But there are also those areas where he knows he could have done better. The mistakes he made are his most important teachers; they tell him the areas in which he lost – and why, what could have been done otherwise, and what he must correct before the next competitive event. The same applies to the expert witness after his day in the judicial arena. A thorough review of his appearance will point out the plusses and minuses of his performance: what he did right, what he did wrong, and how the presentation of testimony could have been changed, improved, or avoided altogether. He must put his mind to work.

The process begins with a "thinking session" encompassing a review of the testimony given and a listing of those points that could have been more constructive or less damaging. After tabulating all items that may have been troublesome, the search should continue with a meeting with his attorney-client. There are two reasons for this post-judicial meeting. The first is to assure the client that the expert gave it his best efforts followed by the desire to assist and, if possible, improve his performance in future cases. The second is to listen to the counsel's comments, both praise and criticism, of the expert's work in the case. This affords the opportunity for review of the entire process from the very beginning through testimony on the witness stand.

If the judicial decision favored his side, the attorney should be pleased and willing to discuss the expert's role constructively. Or conversely, if it went the other way, the attorney may not be

133

Robert J. Crawford

as cordial because of the loss. Nevertheless, he should be
amenable in sharing his thoughts on the expert's contribution. A
good attorney will be willing, time permitting, to explore the
expert's complete role in the case. The talk will bring out the
strengths, as well as any weaknesses, of his involvement -
hopefully followed by suggestions for improvement. The
discussion can also be beneficial to the attorney by the expert
suggesting possible avenues of questioning that were not covered
during the cross-examination phase of adverse experts.

Digging deeper into the review process, addressing each of
the following steps will give thoroughness and direction for the
review:

1. REVIEW OF ORIGINAL SCOPE OF THE
   INVESTIGATION
2. FACTS DISCOVERED DURING THE
   INVESTIGATION PHASE
3. PRESENTATION OF CONCLUSIONS AND
   OPINIONS IN THE REPORT
4. INFORMATION OBTAINED BY READING
   OPPOSING EXPERT'S DEPOSITIONS
5. INFORMATION OBTAINED DURING THE
   DEPOSITION PROCESS
6. CONSULTATION WITH ATTORNEY FOR
   PRESENTATION OF TESTIMONY AT TRIAL
7. IMPRESSIONS OF THE JURY DURING
   TESTIMONY
8. AFTER-TESTIMONY BRAINSTORMING
9. DISCUSSION WITH ATTORNEY-CLIENT ON
   OUTCOME
10. DOCUMENTATION OF ITEMS LEARNED FOR
    FUTURE REFERENCE

1. REVIEW OF ORIGINAL SCOPE OF THE
   INVESTIGATION

A good review must begin with a hard look at the directions given for the investigation of the case. Did the entire scope of study direct its attention to the issues, as presented by the client - or did it become sidetracked by unforeseen developments in the quest for information? On reviewing the history of the study and the expert's role, did he keep on course from beginning to end to attain the goal as requested by the client? Using hindsight, was the attorney-client on track with his assignment? Did he home in on the problem accurately giving the expert positive direction? It is easy to become muddled, confused, and led astray in the avalanche of information that usually comes into the expert's possession; it is surprisingly easy to become disoriented from the original goal. The surest method of overcoming a loss of direction is, at the very beginning, to write down exactly what is required of the expert, noting all specifics. With such a positive beginning, the statement becomes a rudder to keep the study on course. It is also a reliable defense if the expert is faced with a client who subsequently changed from his original direction without the expert's knowledge – as where the case outcome is not the expected goal.

A good example of this is as described in Chapter 5 where I presented a case wherein myself and three other experts in related fields were called to use our collective efforts to provide solid information for the attorney-client side of the issue. As noted therein, he changed tactical directions at trial time without informing any of his experts who had prepared themselves to present their information, as rehearsed for the question/answer direct examination. Even though the expert's prime role is to present his testimony in a completely honest and unbiased manner, he still needs to have a format set with his attorney-client for presenting this evidence at trial; otherwise, those present, including the jury, will not hear it. This attorney lost valuable information and the advantage of credible expert witnesses because of his lack of foresight in reworking the presentation of their testimonies to his new direction.

2. FACTS DISCOVERED DURING THE INVESTIGATION PHASE

The quest for information leads into many areas of study and exploration. With the goal clearly in mind, the road to discovery usually is clear for information gathering. Facts are like gold nuggets in a stream; some are lying there waiting to be discovered. Others often take considerable digging to ferret out. Did the itinerary provide an orderly process for cataloging facts into categories as they affect different aspects of the case? The importance of being able to discern between good, hard evidence and what appears to be evidence cannot be underestimated. It is surprising the number of "important-looking" pieces of information that need to be carefully sifted to glean out the "wheat from the chaff". Information normally comes from three sources: (1) physical evidence from the scene of the problem, (2) documentation in printed or written form, and (3) statements made by witnesses.

In the first, was the expert methodically following his planned itinerary or did he get sidetracked and miss some items? Secondly, did he collect all documented data pertaining to the problem or was something missed by oversight? The oversight may not be by the expert; it may be by the attorney-client because he did not understand the importance of some technical documentation. And third, was he acutely aware that statements made by others may lead into related but different problem areas than the one under study, thus producing confusion? Such are influences by biased witnesses as well as statements made by opposing experts. Did the expert's discovery process uncover extenuating circumstances, facts, or opinions of others that clouded the original goal?

Thus, the importance in having a goal is to keep it firmly in mind while sifting through the information gathered as to what is pertinent and relevant, discarding superfluous information and data. In a recent study, the client informed me of newly discovered information that had significant bearing on the case. After studying the documents, it became apparent that, while the information was related, it was not relevant to the exact goal we were trying to attain. Explanation of why this was so to the client allowed the putting aside of this discovery. Keep your eye on the ball.

3. PRESENTATION OF CONCLUSIONS AND OPINIONS IN THE REPORT

The Report is the document that states the purpose of the investigation, encompassing all items of discovery and analysis, and, presents the expert's opinions regarding the specific issues of the case. Did the Report meet this goal? Was the Report written in clear, understandable English that any person of reasonable intelligence could comprehend? Did it bring out the course of the investigation in a logical and progressive manner that led to the expert's analysis of the facts that formed the basis of his professional opinions? If these requirements have been met, then the Report served its purpose.

The expert must have the ability to express himself in the oral and written word. It is useless for even the most learned practitioner to undertake a study unless he can reduce his thoughts to paper and then present them verbally in court. I have read many expert reports from fields related to my own. They vary from the most simple and concise, conveying their thoughts in layman's language - and a pleasure to read, to others that were so severely lacking in presentation they could easily be improved upon by a fifth grade student. It is appalling to try to read through a maze of improper English, misspelled words, deplorable punctuation and confusing sentences; such misuse greatly reduces the author's credibility in the eyes of the reader.

The same goes for the oral presentation in the question and answer session while on the witness stand. The information has to be understood by the jury. This means that the expert must be able to speak in a clear, audible voice; he must not stumble, mumble, or stutter. With the jury listening to his every word, he must present his answers in common everyday language to be assured they understand what he is saying. If he does otherwise, such as stumbling on his words, the jury will be focused on when he will stumble again and not on the content of what he is saying – that's just human nature.

To have credibility, the Report must be representative of the expert; the expert must present himself as a professional in his line of endeavor. Did the client find the Report easily understood without requiring explanations – other than perhaps some enhancement on a given point? Did the Report meet the criteria? Can it stand alone without the expert's interpretation? Answers to these questions will tell if improvement is needed.

4. INFORMATION OBTAINED BY READING OPPOSING EXPERT'S DEPOSITIONS

To have the opportunity to read and study opposing expert's depositions provides significant insight into both the case and their capabilities. These include personality, educational and training background, licenses, work experience, special education in the issue at question, their study of the case, witnesses with whom they talked, their findings and conclusions, and their opinions to be presented at trial. One of the first questions an expert should ask of his client is for deposition copies of the opposition's experts. In addition to the above noted information, often some new or undiscovered evidence is brought forth that may have been overlooked in the investigation. It is very difficult to be assured that all information has been collected; often evidence disappears and is not available for discovery. At other times, it is simply overlooked and discovered on a second or third visit to the scene. When succeeding visits are not allowed, the opposing expert's deposition will at least afford the opportunity to know of its existence, and some study on the subject information can at least be reviewed.

If opposing experts have not been deposed, this affords the opportunity for the attorney-client to be coached as to questions to be asked. Specifically, to query the extent of the expert's knowledge of the issues in the case, his experience in his field of expertise and in litigative matters, how he is rated as a witness giving testimony, and whether he is newly licensed or a veteran professional. The reading of opposing expert's depositions will reveal whether they are truly good at their vocation or just an over-educated amateur playing the expert game with their "multiple" degrees. Answers to these questions will aid in knowing the opposition.

5. INFORMATION OBTAINED DURING THE DEPOSTION PROCESS

Sometimes an expert can gain vital information during his own deposition. When an opposing attorney prepares for the deposition, part of his homework is to include information gleaned from his own experts on their findings in the case. During the deposition, he will delve into this information, questioning the deposed expert's knowledge of the findings. In this quest to find out what the opposing expert knows, he often brings out bits of information that the deposed expert had not known previously. Thus, being deposed often brings up new information to be studied for its applicability to the case. The deposition also enables the expert to get a "reading" on the opposing attorney and his manner of questioning. At trial, he can expect more of the same – only in much more demanding tones – as the opposing attorney tries to discredit the expert in front of a jury. It is good preparation to know the opposing attorney's tactics for questioning prior to trial. It also is well to remember the attorney is likewise assessing the expert as to the type of witness he will be opposing in the judicial arena. Don't give him any signs of being less than your very best and a witness that thoroughly knows his field and the present issue.

6. CONSULTATION WITH ATTORNEY FOR PRESENTATION OF TESTIMONY AT TRIAL

A major part of the expert's role in litigative cases is to educate the attorney-client. While his expertise lies in the procedures required by the law profession, the content of the case - in technical matters beyond his ken - must be obtained from his experts. Once he understands the story his experts are telling, the information must be put into proper form for the court. Regardless of the type and amount of information brought forth from the experts, it is they - not the attorney - who reveal this information to the court. This only can be done on a question and answer basis. The attorney asks the questions, the expert supplies the answers. It is the expert's job to educate the attorney as to the questions to be asked of him so that he can bring his thoughts, opinions, and conclusions in the case to the court. He is not given the opportunity to "tell his story" except when required to give an explanation to one of his answers. Therefore, the attorney and the expert, working together, compose a "script" of questions and answers that will bring out all information the expert has on the case. The question here is to determine if this was successfully carried out during the trial.

Equally as important is that together they anticipate the opposing counsel's questions expected on cross-examination. A thorough brainstorming, based on all relevant information in the case concerning the expert, is required to cover most of the questions to be asked – and answers to be given. There always will be a few that are missed; so the expert must be up to speed on all information regarding his role in the case. How many times during cross-examination were pertinent questions asked that the expert should have known the answers? Review of what and why will hone skills that will assist in preventing future problems – and embarrassments - in this area.

It must be said, however, not all attorneys are thorough or smart enough to partake in the above exercise. In cases of this sort, the expert only can prepare himself thoroughly and give it his best effort. If possible, it is wise to avoid those who are found to be less than thorough in their work. They may lose the case, but the person with knowledge of the technical aspects of the case was the expert; he lost too.

7. IMPRESSIONS OF THE JURY DURING TESTIMONY

Sitting in the witness box in front of the courtroom can put the calmest expert on edge. All eyes are focused on him as he is questioned and responds with his answers. As previously noted, his demeanor should be one of confidence, answering in a calm, controlled voice. The jury will observe not only what they hear but also what they see. It is imperative that the expert look directly at the jury during his responses; they are the people he is directing his answers to – not the attorney doing the questioning. There are always one or two jurors who have some general knowledge of the expert's field. These people should be singled out – previously by the expert's attorney – and testimony directed to them. They, in turn, will educate other jury members on the meaning of the information; in essence, they are the information leaders of the jury. How they react via their personal demeanor – as well as the remainder of the jury – will give the expert an idea of how he is accepted. Blank stares mean boring; smiles and head nods means understanding and concurrence. It pays to watch the jury's reactions; it gives a clue as to how the expert's performance is going. You also can be assured that the expert's attorney, as well as the opposition's attorney, will be watching these responses. Look at the attorneys when the questions are being asked; look at the jury with the answers.

An example of jury contact was during a trial where I was required to explain some rather technical information in response to specific questions regarding defective workmanship in the construction of a building. All during my presentation, I kept eye contact with those in the jury box who understood what I was explaining. While these two people indicated acknowledgement of their attention with head motions of nodding indicating understanding, the rest of the jury were attentive but not getting the point. At that time, I brought in a model of what I was talking about and went over my previous statements while indicating the various parts of the model to which I was referring. It was interesting to see their eyes become focused, their heads nodding. Some even smiled indicating their understanding. It worked very well as they showed their appreciation for my extra effort to provide this little demonstration; besides, it was fun. The key in trials is the jury; the expert must do everything he can to gain their interest, confidence, and approval of his presentation. It's up to the attorney to make the most of it.

8. AFTER-TESTIMONY BRAINSTORMING

As noted in the beginning of the Chapter, the professional takes time after a contest to review his performance. Within several hours after completing his role as an expert witness at trial, he should spend an hour or two reviewing the case, his performance, the performance of the attorney-client, and the general impression from the jury. He should go over his testimony, his response to questions from both his attorney during direct examination and the opposition's attorney during cross-examination, and any responses from the Judge. For the direct examination, did the "script" play well with all information brought forth as planned? During cross-examination, did the expert handle the queries with ease and confidence? Were there any questions that caused problems? If so, what were they and how could they have been handled differently to alter the outcome. How did the jury respond to his testimony and presence? Did he receive a favorable impression or were they happy to see him dismissed? The parting looks can tell a great deal as to whether they were impressed or glad he is leaving. Finally, the expert should rate himself on his performance. He should have a pretty good idea of how he did and areas that could stand some improvement.

From the above review, constructive ideas will be forthcoming for improving the performance at the next contest. All thoughts generated from this little exercise should be set down for future study in the hours before his next appearance at trial. It will pay dividends in both confidence and poise.

9. DISCUSSION WITH ATTORNEY-CLIENT ON OUTCOME

As soon after the trial as possible, the expert should meet with this attorney-client for a frank discussion of the case. This is a post-mortem of the performance of both parties. Whether the case was won or lost, the reasons why should be brought out for discussion to assist in future work.

If the case was victorious, the judgement will point to the degree. Essentially, the case should be discussed from the beginning; the points of major concern and how they were brought forth during trail – a critique of the "script" on how well it served as well as what could have been added or deleted to be more effective. What points in the case were the most important and what was the emphasis placed upon them. What points proved to be of less use or importance. The interaction between the attorney and the expert in the question/answer process of direct examination should be reviewed; how well it played, and what, if any, charisma was engendered between the two in presenting the story to the jury. From this frank discussion, important information should be forthcoming for the correction of errors made in this case for future reference. The adage "we learn from our mistakes" is never truer than when remembering obviously dumb mistakes made in public forum as in the judicial arena. The discussion is part of the cure for these indiscretions.

10. DOCUMENTATION OF ITEMS LEARNED FOR FUTURE REFERENCE

It is important to keep in mind what the expert is trying to do with this post-trial study. Each of the above steps is to direct the mind to explore his involvement in the case from beginning to end. The object is to gain both insight and experience in honing his skills to improve the performance;

144

improved performance enhances the value of the expert and, accordingly, his reputation and compensation.

Notes from each step should be collected and organized into a notebook depicting two classes of information: (1) items to be reviewed prior to the beginning of each case and especially just prior to trial, and (2) a record of each trial participation to observe and review each performance. The former will be added to the notebook after each trial to enhance the skills gained. The latter gives an overview of the personal growth gained in the progression of past performances.

During this period immediately after a trial, probably the last thing an expert wants to do, emotionally, is to take the time to reflect upon his performance and address the above points. If he feels his presentation was successful and went well, a feeling of satisfaction usually prevails and self-congratulatory thoughts are abundant; perhaps even a little verbal extolling of his performance to the wife and friends is enjoyed. If, on the other hand, his contribution to the effort was less than he had hoped, he would just as soon put the whole episode behind him and move on to more positive challenges.

Regardless of how he feels about the performance, the only way to gain benefit is to take each of the above tasks in hand and glean all information, both positive and negative, from the experience. The object is to hone the skills to ensure the next performance is measurably better than the last, however it went.

CHAPTER 10: A WORD ABOUT FEES

There are two factors that form the basis of successful service business endeavors: (1) to render the appropriate, timely, and expected service to the best of one's ability, and (2) to be paid the agreed fees in a timely manner. Good experts are expensive; as with all other occupations, the amount of compensation depends upon several factors. Perhaps the foremost is the reputation of the consultant. There are many practitioners in the expert witness field, and their capabilities range, like in all occupations, from the very good to the very poor. Reputation is an earned quality; it is gained through performing successful assignments in the eyes of employers as well as those sitting in opposition in the cases.

To illustrate this point personally, I served as an expert witness in a construction defect case that encompassed three successive legal battles – beginning with an arbitration followed by two successive lawsuits - all against the same large residential developer. The procession continued throughout a four-year period. My involvement was to determine what defects - if any - were to be found in a number of upscale residential homes. The defective workmanship, unfortunately for the developer, proved to be true, resulting in the basis for legal action. Approximately six months after the final case was closed, I received a request for expert assistance by the opposing law firm in the previous legal action. It isn't necessarily who wins in a case but the manner in which the expert goes about his task of obtaining and analyzing information and later relating his findings in deposition and court.

Another important factor is the question of service fees. While there are many practitioners of the art and science of expertise, fees should be commensurate with contemporaries who are approximately equal in experience and ability. There will always be extremes at both ends of the spectrum, as some are better than others; but the law of averages applies. High-priced experts who are below standard in performance will eventually experience fewer and fewer assignments. Those who can deliver the expected in an exemplary manner will continue on up the scale. It is well for the expert to keep in touch with the industry to determine if his fees are indeed in line with his contemporaries. These can be determined by friendly acquaintances in the field or by contacting attorney friends for a range of expert fees. Others may be more difficult, as in property owners, real estate brokers, contractors, and developers. They will be quick to recognize and take advantage of an expert who is uninformed as to the prevailing fee rates. Fees vary with the service rendered.

As an example, my fees for inspection and investigative services are at one given hourly rate; for depositions and court appearances, the rates are at a higher hourly rate. The difference is the former requires concentrated thought, data review, thorough digging into the problem, analysis, as well as travel time, etc. However, it pales in comparison to the latter where concentration and recall efforts are required for the grinding questions delivered by the opposition. In court, the questioning becomes even more intense with direct attacks by the opposition in an attempt to render the related information incorrect, inappropriate, and/or to disqualify the expert. The latter obviously deserves higher fees due to maintaining composure under fire, sticking to the facts, and being able to think on one's feet.

Fee discussions begin with the client outlining the scope of the project, locations involved, time lines required for needed information, and persons to contact. Added to this list, the expert should take into consideration expenses that may be incurred in the investigative process. These may take the form of travel, food, and lodging - if the location of the problem is out

147

of the expert's home territory. Consideration also must be given to special equipment that may need to be rented or purchased, engaging specialists in video photography, etc. The client usually will want an idea of estimated costs so that he may plan accordingly – or, in the usual case of an attorney, to obtain approval from his client. I find it to be very beneficial to delay giving a definitive answer until there is time to thoroughly review the tasks involved. If pressed, it is best to give an approximation or range of costs so as to cover unforeseen time or expenses that may be involved. Often the need for additional assistance in the form of consultants in related fields may be required. If an offhand estimate is given at the initial query – and later indications may find it to be low, it will be very difficult to go back for additional fees. Once a quote is given, that is the number the client will remember and expect to pay – and he is usually unhappy about any requested increases – unless adequately justified.

Another important item to keep in mind is the discovery of unanticipated additional work once into the case. When such a matter develops, the client must be told promptly that more work is required than previously thought because of new discovery and that adjustment of fees will be required to complete the process. Don't ever undertake any additional work without previous authorization. One of the quickest ways to alienate a client is to plow onward, pursuing the work, and then bill for the extra cost without his knowledge – even though it is needed and justified. Quite often he will not pay the excess regardless of the extra work required. Get approval first!

While contemplating the fees, it also is important to get a clear understanding of the billing process for the work. Most large firms/offices operate on a monthly cycle for accounts payable. For the client to be able to pay fees, he usually includes the expert's billing along with his own to his client. It is important to obtain information on the client's billing cycle so

148

that the expert's billings may be sent timely for inclusion. Failure to obtain this information may cause a delay in payment. I have missed cycles in the past and have waited up to 90 days for payment – some of it because of carelessness within the client's office, some of it my own fault for not learning the procedure.

The expert has to keep on top of billings to receive timely payments. Small firms or individuals usually operate in a similar manner to the larger firms. However, some clients draw from their own accounts to pay fees. My experience has found many are reluctant to pay until receipt of the second billing – usually 30 days or more after the first - even though they have the funds in their account. It is best to get a clear understanding of the time when fees will be paid and, if delayed, pursue the late payments with diligence. A rather common problem with payment is the "bookkeeping system" used by many businesses, especially the larger firms. This entails the use of "Purchase Orders." Many large firms use this system to keep track of bookkeeping and expenditures. The problem for the expert develops after he has completed his assignment, billed for his services, and waited thirty days or more for payment. On contacting his client, usually after thirty days has passed, he is told the bill was sent to the bookkeeping department and payment is forthcoming. What arrives is a Purchase Order, which must accompany all billings before payment can be made, another thirty-day plus delay.

Another payment problem occurs when, instead of a check, a letter arrives from the client's bookkeeping department noting that payment cannot be made until the expert fills out, signs, and returns an enclosed document giving his Internal Revenue Service Tax Identification Number. In each of these cases, the expert has no option but to comply with the procedures and wait patiently another thirty to sixty days for payment; essentially, back to square one. Obviously, being forewarned is the best way

149

of circumventing this problem. At the outset of the project or case, the expert should ask the client for the billing procedure to get in step with the system and to avoid these needless delays. I have successfully warded off most of the latter problem by including my IRS Identification Number on my billing form.

One of the biggest fallacies in client-expert relationships is to avoid the idea of pressing a client for payment. Many experts feel they don't want to create a problem by bothering a client for payment of fees for fear of losing their business in the future. Whether the client is a steady customer or a one-time stand, it is a mistake not to press past due payments. There are two good reasons for this: first, the idea of not pressing for payment creates an image of the expert as a person who can be intimidated resulting in a loss of respect – as well as late payment. Secondly, working with such a client is wasting time that could be used more effectively used serving another who does pay on time. BEWARE of clients who make collection a problem!

There are times when additional services are required after the initial work has been completed. Fees for any extra work usually are provided for in the original letter of agreement. It is important, however, to inform the client of any forthcoming increases in the fee schedule as of a given date. If not so informed, he may properly assume the same fee structure is applicable and may question or even protest any increases after the fact. If a new case develops out of the previous one, as often happens, it is best to originate a new letter of agreement with tasks and fees noted accordingly.

In every work assignment, the expert should draft a Letter of Agreement. It should contain the essential information as to the requesting party, statement of the problem, services required, specific tasks to be performed, time required, fees for the given services, timely payment upon billings, and the payment of litigation costs if the fees are not forthcoming. It is good practice to obtain a significant retainer at the outset, ranging from 10% to 50% of the expected fees, depending upon the anticipated integrity of the client – long-standing clients may, of course, be excepted. The Agreement should be signed and dated by both parties. While some may prefer a full-blown contract with whereas's and wherefore's, I find this brief form has served me well without problems for many years. Keep agreements simple, clear, and easy to understand.

Fees make take various forms in method of payment. As stated above, I usually base my fees on time and material; i.e., a given hourly amount for the various tasks, expenses, materials, secretarial, etc. These are the norm for so-called "open-ended" services when I am not able to estimate the time involved. This is why the "approximate range of fees" proposal works. Fee discussions begin with the presentation of my fee schedule to the client during our first meeting. As stated above, it is extremely important to bring up the subject of additional work that may be required beyond the original scope and to obtain an understanding of how fees will be structured in that event. There are some cases when it is clear as to the time involved to perform a specific service. On these, I can accurately estimate the fees and quote a lump sum. However, I always put in the original Agreement the scope services involved. If subsequent additional work is required, it is then specifically identified – differentiated from the original services – and quoted on a time and material basis.

Several forms of fee compensation are the lump sum, the hourly rate, the daily rate, and percentage-of-award – if applicable. The lump sum, as described above, is a fee quote to do an entire project or case. For the work as described in the Agreement, it does two things: (1) the client knows exactly what he will have to pay, and (2) the expert knows precisely what fees will be forthcoming. This process works well on very small, easily determined cases where the expert's involvement is minor – as in a few hour's work on research or an opinion on an issue. It is disadvantageous in mid-size to large cases because of the unknown time and expenses involved – and it often works in both directions.

An example is a case where an architect charged a considerable lump sum fee to do a structural investigation of alleged defects in a rather large group of residences. This was desirable to his attorney-client as he knew in advance precisely what fees would be paid for the information. Unfortunately, the architect's investigation program amounted to inspecting about 15% of the total number of houses involved spending less than an hour per house. Further, he wrote a minimal report on his findings, engaged in several hours of explanation with his counsel, and testified for several hours in court. His considered opinion indicated there were no problems with the buildings. Observation from outside looking in, it was obvious that he did not do an effective job of investigating, was of ineffective assistance in court, and grossly overcharged his clients for the work. His testimonial conclusions were proven to be incorrect in light of opposing documented evidence. A poor selection by his attorney-clients; they would have paid considerably less had the rate been on an hourly basis. The lump sum fee also may work in reverse, however, as when the expert underestimates his fees and becomes unhappy for having to complete more work than anticipated; this often leads to hurrying and missing important data in the discovery process.

The daily rate is similar to the hourly rate. The only difference is the daily rate is based upon the expert working on the case one day at a time – all day, each day. This means he will work a required number of days – often sequentially – until the task is complete. An example would be if the expert traveled to a distant place to do his investigative work. Upon return, he would spend the number of days required to complete the process. Fees are on a daily basis, but usually figured as an hourly rate multiplied by the usual eight-hour day. I have seen very few fees structured on this premise.

Attorneys normally work on either an hourly rate or a percentage-of-award basis, sometimes both, depending upon the client and the situation. Most, from what I have seen, opt for the award basis on mid-size to large cases as they usually command fees of 30% to 40% of the proceeds; certainly significantly higher than what could be gained on an hourly rate – provided, of course, that they win. Obviously, they are taking a calculated risk of not being paid for their services. For an expert witness to base his fees on a percentage-of-award would be a clear indication of a bias involvement. He would hope to gain by affecting a favorable award verdict on his client's behalf. A major prerequisite of an expert witness is to be unbiased; i.e., to give an honest opinion of the issues involved in a case. To do otherwise, or have an outside interest in the outcome, would be a disservice to his client and the court and provide grounds for impeachment. I have heard talk among experts discussing this method; however, they are probably just reflecting upon the difference in fees between those received by the attorneys and their own.

It is important for billing forms to contain all information required to identify the case involved. It should also be kept simple, easy to read, and follow the process of fees billed and payments made. The form I use contains the following:

1. Title Page. This is simply the name, address, and other pertinent data pertaining to my office.
2. Date of billing. Simply the date on which the billing was sent from my office for payment.
3. Job Number. All projects, cases, etc., in my office are given a number for identification purposes. For convenience, I use the current year along with the

specific number for the individual project. As an example: 153-98. The first number is the specific job, the second is the year in which the job was contracted. I begin each year with the number 101 and run them consecutively through the year. This is a simple system which works very well in separating the various projects.

4. The client's name and title, firm name and address.
5. The latest date on which work was done.
6. The name, address, and client's job or file number for identification with their records for the case or project.
7. The listing of the services provided in accordance with the Agreement.
8. Description of fees due (numerically): (1) if an hourly billing, it would be the hours spent multiplied by the hourly rate with the resulting total; or (2) the amount due per the Agreement as in a lump sum billing or percentage completed thereof.
9. Amount due per prior billing and the date thereof.
10. The payments received on the account since the last billing.
11. The resulting total from the above prior amount due less payments.
12. The new fees due at this billing.
13. The total fees due from prior and current billing.
14. My IRS Tax Identification Number.
15. A special "Thank You" for the opportunity to work with them and for the payment.
16. At the bottom, a statement that all fees are due upon receipt of billing; fees past due by 30 days will be subject to a service charge of 1.5% per month.

This simple one page sheet contains all of the information necessary to describe the billing, to whom it is directed, the cutoff date for billing purposes, a description of the project and services involved along with current charges followed by a numerical calculation of the billing and fees due. The final statement of charging a 1.5% fee is termed a "service charge" for administrative purposes. The original is mailed to the client and a second copy is kept in the "accounts receivable" file. Upon receipt of the fees, the file copy is then transferred to the "accounts paid" file. In working with clients, the above information was developed into the current form to avoid misunderstandings by clearly pointing out fees due and past payments. As subsequent billings are sent, only the billing date and amounts change duly reflecting fees received in the interim since the previous billing - unless, of course, additional work was performed. This system also allows keeping abreast of fee payments at a glance.

One of the most important jobs in an office is to see that billings are promptly sent out containing the correct information. Billing for services are tallied and sent out initially within two days upon conclusion of a project. Subsequent billings are usually mailed on the 25th day of each month – or earlier - to coincide with the clients timely billing cycle; most bill on the following month's first business day.

Records are kept in the above referenced "accounts" files. The files serve several purposes. First, they are a track record of the current and past jobs. From these, it is relatively simple to keep track of fee payments and the amounts charged for each job. It also gives a historical perspective of fees charged over time and information as to whether current charges are commensurate with contemporaries in the field or, if justified for other reasons, increases are reasonable. Finally, they are a back-up record for income tax information.

The collection of past due fees is an important matter. There are several ways that may be used once all avenues of personal effort have been exhausted. If the bills are considered small, the simplest method is through Small Claims Court. Most slow pay clients, especially attorneys, will come through upon receipt of a Small Claims Court subpoena. Another method is through the use of a collection agency. These people are usually successful if the slow payer has the funds; if not, they will make a few tries – then give up. Collection agency fees vary – some high, some low - but part payment is better than no payment. Beyond these are the lawyers. Collections for significantly larger sums are usually made by attorneys specializing in the practice. Their fees may be either on an hourly basis or a percentage of the funds collected. Whichever way selected, they are usually successful if the nonpayer has funds. If the nonpaying client does not have the funds for payment, perhaps the best method of collecting is to have him sign a note for the amount and make installment payments. I have used this successfully a number of times. They are not always prompt and may need some encouragement for past due payments, but eventually they pay because of the note. The option to go to court is always open if the amount due is worth pursuing.

In summary, the collection of monies earned is perhaps the most important factor in any business. This is the reason why we undertake the assignments that come to us. Obviously, the enjoyment to engage in and perform the work one is educated and trained to do has its own rewards; but the funds received from our efforts provide us with our livelihood. It is difficult to meet our financial obligations without them. In negotiating with clients, remember the service business is a two-way street. The client wants the information the expert can produce; the expert wants the income to pay his expenses and operate a successful business. There are many pitfalls to be avoided in the successful pursuit of one's work. These range from the initial client agreements to the collection of the fees. Having the know-how to avoid undesirable circumstances goes a long way to making one's work much more entertaining and enjoyable.

As a final point in the management of the fees, there is the identification problem of expert witnesses versus percipient witnesses. As defined, an Expert Witness is a person who, by virtue of their education, training and/or experience, is qualified to testify under oath in legal proceedings as to their opinion of an issue in their field of knowledge. A Percipient Witness is a person who may have specific knowledge of the field at issue but is used as a witness to items that may only be generally related to his specific knowledge and not used for his professional opinion of the issue of concern.

The difference is considerable with respect to the compensation for appearing as a witness. There are several governmental code sections that specifically relate to this situation. The essence of them state that an expert witness is entitled to his usual and customary fees for attending legal appearances and rendering his expert opinions. The percipient witness, being called as other than an expert, is entitled to only a predetermined sum for his appearance and testimony and is normally considerably less than that of the expert.

This is an issue that should be discussed with your attorney if the case is involved in a legal proceeding. It should be clearly understood by all parties that your appearance is as an expert witness commensurate with your standard fee schedule. As an example, I was deposed as an expert witness in a case where I was the prime consultant for my attorney-client. After some two hours of deposition tactics, the opposing attorney – with an ingratiating smile – thanked me for my cooperation and presented a check for approximately one-fourth of what I normally charge for my time. It was certainly a learning experience. My attorney-client corrected the situation upon my request.

Another issue that should be brought forth is the timeliness of payment for depositions. Normally, an expert will send a bill to the opposing attorney within a day or two of the proceedings, as is customary because he called the deposition. Most will pay within a few days. However, I had one attorney who, upon receipt of my billing, informed me that I would be paid after the case was settled. A little legal research found Code Section CCP Paragraph 2037.7(b), stating that expert witnesses are to be paid within five days upon receipt of their billing – unless certain circumstances allow for a delay. Upon politely enlightening the deposing attorney with this news, I received prompt payment.

There are a number of obstacles to be overcome in the legal arena that affect expert witnesses. Most expert witnesses will learn them by experience. The purpose here is to share what I have learned during my years of practice in hopes that for others the wheel will not have to be reinvented.

GLOSSARY OF LEGAL TERMS.

Listed herewith are definitions of some of the legal terms used in the field of law by attorneys who work with expert witnesses. They are included here for quick reference when encountered in the course of work. The use of Latin is prevalent in the law as in medicine. Where these words and phrases occur, they are so described with their literal translation. If more elaborate meanings or precise pronunciations are required, I recommend any of the legal dictionaries available.

AB INITIO: Latin – from the beginning.

ACTION: A condition whereby the first party litigates against a second party for wrongdoing or harm.

AD HOC: Latin – a condition that exists for a specific purpose only, such as a committee charged with a given purpose.

AMICUS CURIAE: Latin – Literally means "a friend of the court;" one who assists the court in its duties.

ANSWER: The specific response of a defendant to the charge brought against him by the plaintiff.

A POSTERIORI: Latin – Literally meaning "from the most recent viewpoint."

A PRIORI: Latin – Literally meaning "from the beginning, from the first."

ARBITER: A court appointee entitled to decide a legal, controversial issue.

ARBITRATION: A legal proceeding in which the issues on both sides are brought forth with a judgement rendered by the presiding officer, the Arbitrator.

ARBITRATOR: A knowledgeable person(s) selected by each side in a dispute to hear and render judgement in the matter.

ATTEST: The act of bearing witness with a signature declaring an issue or document is true.

BONA FIDE: Latin – Literally means "in good faith," not false.

BRIEF: A document presenting the facts of a case to the court in precise legal terms.

CASE: In law, a lawsuit or action to be filed with the court.

CAVEAT: Latin – Literally means "let him beware," a warning.

CAVEAT EMPTOR: Latin – Literally means "let the buyer beware" or warned.

CERTIORARI: Latin – Literally means "to be informed of". A methodology for application to a higher court for a case.

CHALLENGE: The right of a litigant attorney in a case to cause the dismissal of a juror without cause.

CIVIL: That branch of law that pertains to legal actions other than criminal in nature.

COMPETENT: Being capable of rational and prudent thought, as well versed in the subject under consideration.

CONCUR: To be in agreement with.

CONFLICT OF INTEREST: A condition where a party to one side of a lawsuit has an underlying interest in the opposing side thereby causing his opinion to be biased.

CONTINGENT FEE: A fee paid to a lawyer as a retainer in a case with recovery by the client dependent upon its outcome.

CONTRA: Latin – Literally means "against, in opposition to."

CORPUS: Latin – Literally means "the body," the essence of an issue or thing.

COUNSEL: An attorney who is a legal advisor.

COURT: Federal law establishes the court as the institution for settlement of legal disputes among its citizens.

COURT CALENDAR: The schedule containing the list of cases due to be presented in a specific court.

COVENENT: Agreement to a formal contract or obligation.

CROSS-EXAMINATION: The querying of a witness by an opposing attorney while under oath in deposition or court.

CULPABLE: Being wrong for acting without regard for another causing harm and thereby deserving of punishment.

DAMAGES: A monetary award by the court for losses incurred to a plaintiff by an act on the defendant's part.

DECEIT: The deliberate misrepresentation of a statement or fact upon which another relies as the truth; a duping of another who acts in good faith.

DE FACTO: Latin – Literally means "in fact;" a condition of operation with the appearance of being authorized, but not officially legal.

DEFAULT JUDGEMENT: A judgement rendered against a defendant due to non-appearance in court.

DEFECTIVE: Not complete, faulty; having parts that are not in accordance with specifications.

DEFENDANT: The party that is in receipt of a lawsuit; the one who a legal claim is filed against for damages.

DEFENSE: The plea entered into court by a defendant refuting the claims filed against him by the plaintiff.

DEFRAUD: To deceive a person thereby depriving him of property or interest therein by fraud.

DELIBERATION: The process whereby jurors determine the outcome of a case and render judgement.

DEMURRER: A formal charge whereby the facts alleged in a case do not contain enough evidence of wrongdoing for it to proceed into litigation.

DENIAL: The act of denying the validity of the facts in a case that charge the defendant with wrongdoing stating they do not contain sufficient information for prosecution.

DEPONENT: In the discovery process, a person called into deposition for the purpose of eliciting information and opinions, as an expert witness.

DEPOSE: The act of questioning a witness during deposition.

DISBAR: The act of relieving an attorney of his license to practice law due to improper actions regarding behavior.

DISCHARGE: To be relieved of an obligation of an act or debt.

DISCONTINUANCE: During a court proceeding, the plaintiff's attorney voluntarily calls for an end to the action.

DISCOVERY: The process whereby attorneys question opposing witnesses for information they possess that is pertinent to the case.

DISCRETION: The freedom an officer of the court enjoys to chose a course of action within the legal bounds placed upon him.

DISMISS: To voluntarily terminate a proceeding during any part of the on-going process.

DISMISSAL: A cancellation of a motion made before the court, a denial of the motion.

DISPOSITION: A meeting out of the proceeds or components of an estate or other collection of items.

DISPOSSESS: To legally bring about a departure of another from a place or location; an eviction.

DISTRESS: A process whereby one person seizes the property of another without proper legal authority in the satisfaction of an indebtedness or claim.

DOCKET: The list of upcoming cases as shown on the court's schedule.

DOMICILE: The permanent or legal residence of a person.

DONOR: A person who freely presents another with a gift without any expectation of return.

DUE CARE: Care offered and given to a recipient that a prudent person would render under the conditions which it was bestowed.

DURESS: The act of requiring action of a person under a control that such would not otherwise be freely given.

EARNEST: As in money; something that has value acting as collateral for payment of something in return.

EMINENT DOMAIN: The taking of private property by the government for the public's use with the owner be justly compensated for its market value.

ENCROACHMENT: To enter onto another's property without their permission, as in building a structure across the property line onto another's property.

ENCUMBRANCE: A right of others to access onto an owner's property without his permission, as in a roadway or utility easement.

EN GROS: From the French meaning a gross amount of substance or goods.

EQUITABLE: That what is described as fair and just with due compensation as in an exchange of goods or property.

ERGO: Latin – Literally means "therefore" or "thus."

ERROR: The making of a mistake or the loss of being accurate in the subject under discussion or belief.

ESQUIRE: A special title that is reserved for the legal profession; lawyers, attorneys.

ESTOPPEL: A barrier against a statement or act that would produce an unjust decision.

ET AL: Latin – Literally meaning "and others."

ET NON: Latin – Literally meaning "and not."

EVIDENCE: Information gathered to support or deny a claim that is presented in a judicial proceeding.

EX GRATIA: A deed or action that is done out of generosity rather than as a paid task.

EX OFFICIO: Latin – A person who becomes a member of a group, board or body because of the office or post he holds.

EXPERT WITNESS: A person who, due to his knowledge, experience and expertise is qualified to render an opinion on the issue under consideration in a judicial proceeding.

EXPERT TESTIMONY: Statements made in a judicial proceeding by a person who is qualified to render such opinions on the issue under consideration.

EX POST FACTO: Latin - Literally meaning "after the fact;" a situation where an act committed during the time preceding the passage of a law that made such an act illegal.

EXTRAJUDICIAL: An act or statement made that was not directly connected to a legal proceeding, as in a confession to others of a crime.

EYEWITNESS: A person who was present and observed personally an act and can testify to such.

FACT: The existence of a condition, material or substance that is established by evidence.

FACTO: Latin – Literally meaning "in fact."

FAULT: An error or mistake in reference to a person's conduct in performance; in materials, it means a "defect" in the make-up, manufacture or production of an item.

FIDUCIARY: A duty required of a person to perform obligatory tasks in accord with the functions of the office he holds.

FORENSIC: Refers to the application of specialized skills that are performed by knowledgeable persons on behalf of the judicial system as in courts of law.

FORFEITURE: The loss of money or property on a permanent basis for failure to comply with the legal aspects of an agreement or contract.

FRIVOLOUS: An action or claim that cannot be defended or prosecuted in a court of law.

FULL DISCLOSURE: The requirement to disclose all facts, details and conditions of an action or occurrence including all parts and portions of the issue.

GAG ORDER: The silencing by the court restricting comments on an issue before the court.

GARNISHMENT: The attachment of money by the court due to the defendant from the plaintiff and placed in the possession of a third party until resolution of the issue by the court.

GOOD FAITH: The intention to honestly meet an obligation.

GRATIS: Latin – Literally meaning "freely given;" goods or service given without expectation of return value.

HEARING: A legal proceeding where evidential testimony is given concerning an issue and a decision is rendered based upon the evidence presented.

IBID: Latin – Literally meaning "in the same place or manner;" commonly used to refer to the same source of a previously described reference.

IMMATERIAL: Information that is not pertinent or significant to the issue.

IMPEACH: With respect to expert witnesses, a process to challenge the truthfulness or bias of a witness while giving testimony under oath.

IN ABSENTIA: Latin – Literally meaning "in absence," as in the situation where an expert may be unable to appear in court and his deposition is read as his testimony "In Absentia".

INCOMPETENCY: The inability to present information due to lack of skill, knowledge or expertise in a legal proceeding; it may also be due to illness or loss of physical or mental faculties.

IN DELICTO: Latin – Literally means "at fault."

INDUSTRY STANDARDS: The normal skill and competence used in the performance of duties that is acceptable as the standard throughout the industry under consideration.

IN FUTURO: Latin – Literally meaning "in the future."

IN GENERE: Latin – Literally meaning "in kind" as in the same type or class.

INGRESS AND EGRESS: Entering into or leaving from an area or space; a building.

IN HOC: Latin – Literally meaning "in this;" as in this instance or case.

IN PERPETUITY: Latin – Literally meaning "existing forever."

IN RE: Latin – Literally meaning "in regard to or in reply to."

INTERROGATORIES: A legal tool used before trial in civil cases for discovery purposes by submitting written questions to adversarial parties on the issue who must answer under oath.

IN TOTO: Latin – Literally meaning "in total."

IPSO FACTO: Latin – Literally meaning "by the fact in and of itself."

IPSO JURE: Latin – Literally meaning "by the law itself."

IRRELEVANT: Testimony given that does not pertain to the issue; information not pertinent to the case.

ISSUE: A case or point of discussion in a court of law.

JUDGE: The person who presides at a trial in a court of law.

JUDGEMENT: The result in the settlement of a dispute in a court of law.

JURAT: Latin – Literally means "has been sworn."

JURISDICTION: The geographical territory encompassed by the local governing authority to hear and determine legal cases.

JURIST: A judge; one who is schooled in the law.

JUROR: A person selected for jury duty.

JURY: A group of twelve elected jurors to serve at trial to decide on the facts presented.

JURY TRIAL: A proceeding wherein the jury decides the issue of fact in a court of law.
KANGAROO COURT: A non-legal proceeding that has no governmental authority to pass judgement or render damages on persons connected thereto.

LATENT DEFECT: A defect in the composition, manufacture or construction of an item not discoverable by knowledge, appearance, or reasonable care or usage.

Robert J. Crawford

LAW: That body of rules governing the actions of society established by the legislative body having jurisdiction over the geographical area.

LEADING QUESTION: A question posed to a witness in such a manner that it points to the answer that is expected; leading the witness as to what he is expected to say.

LEASE: A document of agreement between an owner (lessor) of a property and a user (lessee) granting possession of the property for a period of time and a given consideration (value) subject to the terms of the agreement.

LEVY: The ability to assess or collect a sum of money against a property as in governmental taxing.

LIABILITY: An obligation of the liable party to perform as required by law, agreement, or obligation.

LIABLE: An obligation under law; to be responsible.

LIBEL: Usually false printed or photographic material that defames a living person or darkens the reputation of a deceased person.

LICENSE: A legal document usually issued by a governmental authority that authorizes the performance of particular tasks or duties.

LIEN: A monetary fee or its equivalent against a property used as a form of security for the performance of a specific task or debt.

LIQUIDATED DAMAGES: An agreed upon monetary amount to be paid as damages to one party of a contract as restitution if the other party fails to perform per the agreement.

LIS PENDENS: Latin – Literally meaning "a pending lawsuit."

LITIGATION: A contest to determine the enforcement of legal rights in a court of law.

LOCUS: Latin – Literally meaning "the place."

LOITER: The act of aimlessly occupying a given public place for no apparent reason.

MAGISTRATE: A name usually reserved for lesser judicial officials.

MALFEASANCE: The committing of an illegal or wrongful act, especially the interference of an officer performing duty.

MALICE: A state of mind that causes a person to commit an unlawful act against another without just cause.

MALPRACTICE: Improper execution of duties as defined by the rules under which the perpetrator practices.

MANDATE: An order by a judge or court.

MARKET VALUE: The value of an object or property as determined by the free market.

MEDIATION: A manner of settling disputes by a meeting of opposing parties to come to an acceptable agreement usually officiated by an attorney.

MINUTES: A written record of a legal proceeding.

MISNOMER: A mistake made in a person's name; citing the wrong name of a person.

MODUS OPERANDI: Latin – Literally means "the method of operation."

MORATORIUM: A temporary halting of a proceeding; a delay in the process.

NEGLIGENCE: The failure to use reasonable care in the execution of a task that another person with the same qualifications would normally use.

NEGOTIATION: A process wherein parties to a contest meet to come to an agreement to settle a dispute.

NOLO CONTENDERE: Latin – Literally means "I do not wish to contend or fight."

NONFEASANCE: Failure of an agent or representative of a principal to perform a duty that he has agreed to perform.

NON SEQUITUR: Latin – Literally means "it does not follow."

NOTA BENE: Latin – Literally means "note well."

NOTARY PUBLIC: An officer licensed by the government to administer oaths and to certify persons and their signatures to documents as true.

NOTE: A written document that promises to repay a given debt in accord with the terms of the agreement.

NUGATORY: A judicial proceeding in a court of law that lacks jurisdiction.

OATH: A verbal obligation to tell the truth in a judicial proceeding or a statement made to be true.

OBJECT: A verbal response by an attorney in a court proceeding wherein the information offered is improper or not relevant to the issue and asks the judge to have it removed from the record.

OF COUNSEL: Attorneys consulting with the principal attorney on a case.

OMISSION: A duty required by law that has not been done; failure to do as required.

OPINION: Statement of the reason for a court's judgement.

ORDINANCE: A law that is issued by local governmental authority, as a city or county.

PAR: An equality of established value; not the actual market price.

PARALEGAL: A person performing office duties associated with the practice of law but not a member of the Bar.

PATENT DEFECT: A defect easily recognized in an item upon a reasonably scrutinized inspection.

PECUNIARY: Money or its equivalent in goods or services.

PER ANNUM: Latin – Literally means "per year."

PER CAPITA: Latin – Literally means "by the head."

PERCIPIENT WITNESS – A person who may have specific knowledge of the field at issue, but is used as a witness to items that may only be generally related to his specific knowledge and is not used for his professional opinion of the issue of concern.

PER DIEM: Latin – Literally means "by the day."

PER QUOD: Latin – Literally means "through which."

PER SE: Latin – Literally means "through itself."

PLAGIARISM: The adoption of literary work as one's own without appropriate permission; a form of theft.

PLAINTIFF: The initiator of legal action.

POLYGRAPH: A mechanical recording devise commonly known as a "lie detector."

POWER OF ATTORNEY: A written document that allows a principal to authorize another person as his agent to act on his behalf in specific legal issues.

PRIMA FACIE: Latin – Literally means "at first view;" on its face.

PRIVILEGE: A position that allows a person, business or group of people to have advantages not shared by the general business or social community.

PRIVILEGED COMMUNICATION: Oral and written communications in the legal community that have confidentiality.

PRO BONO PUBLICO: Latin – Literally means "for public good or welfare."

PROCEDURE: The course of legal progression in a lawsuit.

PRO FORMA: Latin – Literally means "for the sake of form."

PRO RATA: Latin – Literally means "according to the rate."

PRO SE: Latin – Literally means "for himself."

PRO TEM: Latin – Literally means "for the time being."

PROVISIO: A condition to prevent misunderstanding.

QUASH: A tactic to oust a condition by judicial action.

QUASI: Latin – Literally means "as it were."

QUID PRO QUO: Latin – Literally means "something for something."

RATIO DECEDENDI: Latin – Literally means "the reason for the decision."

RATIO LEGIS: Latin – Literally means "legal reasoning."

READY WILLING AND ABLE: Readily prepared; willing and able to act.

REASONABLE PERSON: An intelligent person who is capable of initiating thought, judgement, and is reasonable in his thinking processes to arrive at a reasonable conclusion based upon information received.

REBUTTAL EVIDENCE: The presentation of evidence that counters information presented by the opposing side in a legal contest.

RECORD: The written preservation of information given during testimony in a legal proceeding.

RECOVERY: The monetary amount recovered or collected pursuant to a judgement.

RELEVANCY: The testing of evidence in a legal proceeding to determine if it is admissible.

RES: Latin – Literally means "a thing."

RESCIND: To revoke a contract returning the participants to their original positions.

RESCISSION: To void or cancel a contract thereby returning the contracting parties to their original pre-agreement positions.

RES GESTAE: Latin – Literally means "the thing done."

RES JUDICATA: Latin – Literally means "a thing decided."

RESPITE: Essentially a delay or postponement.

RETAINER: A sum of money paid to an attorney to begin representation in a lawsuit.

REVOKE: To rescind or cancel an authority previously given.

SETTLEMENT: A conclusion resulting in an agreed compensation by both parties in a lawsuit.

SINE DIE: Latin –Literally means "without day, without time."

SINE QUA NON: Latin – Literally means "without which not" the essence of a thing.

SITUS: The legal location of a place or thing under consideration.

STANDARD OF CARE: The skill that a qualified person would use to meet the minimum requirements for industry standards for the work in which he is engaged.

STATUS QUO: The way in which conditions exist.

STATUTE: A rule of law as enacted by the governing governmental body.

STATUTE OF LIMITATIONS: That time period in which an action must be brought against a wrongdoer before expiration.

STIPULATION: In a matter in a legal proceeding, it is an agreement on an item by both sides in the case.

SUB-CONTRACTOR: A specialty contractor working under contract to a general contractor to ply his special trade on a project.
SUB JUDICE: Latin – Literally means "under a court."

SUB MODO: Latin – Literally means "under a qualification."

SUB NOMINE: Latin – Literally means "under the name."

SUBPOENA: A legal document issued by a court requiring the appearance of the named person; sometimes includes all files and records on a case and is referred to as a subpoena duces tecum.

SUB SILENTIO: Latin – Literally means "under silence."

SUI GENERIS: Latin – Literally means "of its own kind;" in its own class.

SUI JURIS: Latin – Literally means "of his own right;" one who attains majority.

SUIT: A legal proceeding to provide remedy for a possible injustice.

SUMMONS: A court order requiring the appearance of a defendant.

SUO NOMINE: Latin – Literally meaning "in his own name."

SUPRA: Latin – Literally meaning "above;" as in the "above text."

TESTIMONY: A statement or statements made by a witness under oath in a legal proceeding.

TORT: Essentially a wrongdoing under civil law that is a breach of legal duty in the society in which one lives.

TRIAL: An established court proceeding in which an issue is tested by before a jury of peers with a judgement rendered at the conclusion.

VACATE: To set to one side, to render void or vacant.

VEL NON: Latin – Literally means "or not."

VENIRE: Latin – Literally means "to come;" a summons of jurors to a case.

VERDICT: The decision rendered by a jury in a court proceeding.

VOIR DIRE: Latin – Literally means "to speak the truth."

WITNESS: One who gives testimony under oath in a legal proceeding.

WORK PRODUCT: Work done by an attorney in the preparation and presentation of a case in representing his client that is not subject to discovery.

WRIT: A legal document issued under court authority compelling a person to perform the duties specified therein.

Robert J. Crawford

About the Author

Robert J. Crawford has been in engaged in private practice as a civil and structural engineer designing buildings for 40 years. He has served as an expert witness in litigation activities for over 25 years and presently continues his efforts in both areas.